UNTIL DARWIN, SCIENCE, HUMAN VARIETY AND THE ORIGINS OF RACE

UNTIL DARWIN, SCIENCE, HUMAN VARIETY AND THE ORIGINS OF RACE

BY

B. Ricardo Brown

Routledge
Taylor & Francis Group

LONDON AND NEW YORK

First published 2010 by Pickering & Chatto (Publishers) Limited

Published 2016 by Routledge
2 Park Square, Milton Park, Abingdon, Oxfordshire OX14 4RN
711 Third Avenue, New York, NY 10017, USA

First issued in paperback 2015

Routledge is an imprint of the Taylor & Francis Group, an informa business

BRITISH LIBRARY CATALOGUING IN PUBLICATION DATA

Brown, B. Ricardo.
Until Darwin, science, human variety and the origins of race.
1. Human evolution – Research – History. 2. Monogenism and polygenism.
3. Race relations – History. 4. Science – Social aspects.
I. Title
599.9'38'072-dc22

ISBN-13: 978-1-138-66144-8 (pbk)
ISBN-13: 978-1-8489-3100-8 (hbk)

Typeset by Pickering & Chatto (Publishers) Limited

CONTENTS

... and not least we did this for those who are called 'foreigners', though they are not really so. For, while the various segments of the earth give different people a different country, the whole compass of this world gives all people a single country, the entire earth, and a single home, the world.

Diogenes of Oinoanda, The Epicurean Inscription

ACKNOWLEDGEMENTS

This author never wanted to write on the subject of 'race' and certainly never to write a monograph on the topic. Scattered references, occasional plays on words, clever juxtapositions of words and things should have been enough of a commentary, as if a commentary was necessary. Unfortunately, even with all the forests that have been sacrificed for the publication of works on the subject, very few works address the history or genealogy of race as either a concept or as something akin to a scientific ideology. Certainly a very few did; works like those by Stanton, Gossett and Gould did place race in its proper context, but the vast majority of works on race begin with the assumption that race is not a concept but a physical manifestation of a biological value that overdetermines the course of one's life.

In opposition to the tendency to 'notice race', this work does not attempt to 'explain racial differences' nor is it an attempt to debunk the concept of race because race has been debunked so many times before and yet it continues to exert a hold on the Western imagination. Instead, it is about the ways and systems of classification that we have used in order to create, in terms of the production of knowledge, a metaphysical construction of human bodies. Knowledge of these systems of classification allows the possibility for debunking, but it is only a possibility in so far as the readers of this text might imagine for themselves the possibilities and terrors that emerged from our own everyday life.

This is an introductory essay to a larger project. At some time in the future, additional work will more fully explore ancient scientific discussions of human variety and the period of the rise of sociology and theories of degeneracy. If this manuscript is filled with the names of the obscure and forgotten except by those who take an obscene interest in wretched knowledge, then remember the comment by Michael Wood, who said that in the end only the greatest villains are remembered and the everyday criminals – the proletariat of domination – quickly recede from memory. To this, one might surely add that, in their own way, the crimes of the obscure are left to be eternally repeated just like the notions of common sense they believed they embodied.

My great thanks to those who have endured my engagement in this project over the past twelve years. Nancy K. Cauthen, Stanley Aronowitz, Susan Buck-Morss, Patricia T. Clough, Jim Monsonis, Ivan Zatz, Melinda Russell, Kenneth Cauthen, David Byrd, Frederick Daley, Robert Barash, Suzanne Verderber and Amy Palmisano provided comments and support along the all too long way to the completion of this manuscript. The generous assistance of the librarians of the American Museum of Natural History who nicely tolerated my 'occasional' all day visits. Julie Wilson of Pickering & Chatto for her attentive, precise and painstaking editing. The members of the Situations collective, my supportive colleagues at Pratt Institute and the participants in the 2004 ASA panel, who reminded me that no matter how obscure one's work, no one actually works alone. My wonderful nurses and the best team of doctors, Greg Riley, Kenneth Rosenzweig and Joshua Sonnet, who saw me through my a terrible year of thymoma just as I was finishing this work.

Of course, how my family tolerates my indulgence in all this wretched knowledge will always amaze me. My companion and wife, Nancy, has always supported my work, so I give this work to her and our sons Jacob and Evan because it is really for them, my sister Renee and our late parents Lillian Harllee and Bennie Brown. I think they wanted me one day to write a book like this.

LIST OF FIGURES AND TABLES

For
Jacob H. and Evan F. Cauthen-Brown

INTRODUCTION: *ECCE HOMO* OR SLAVERY AND HUMAN VARIETY

The history of science is a history of forgetting. It is the history of how scientific truth emerges from the murky cacophony of words and things that were once said and built, but are now silenced and buried. At the moment a regime of scientific truth coalesces, this cacophony is enveloped within a rational, ordered and yet arbitrary universal system. But we should pause to remember that the elements of this system were already present in the anarchy it replaced. For reasons of practicality, we are taught to forget the chaos which preceded contemporary knowledge. At the same time, those elements of wretched knowledge that we thought were finally repressed by truth continue to emerge over and over again. In 1999, for example, most of the sociologists and anthropologists in the United States and Canada received in the mail an edited version of a 300-page work purporting to prove the inferiority of blacks and Asians relative to whites. The book was by a tenured professor at a respected Canadian university and published by a reputable press associated with a major American university.

On the other hand, it is certainly true that many insightful critiques of the concept of race have already been produced. The best of these works carry on the tradition of examining race not as an essential aspect of bodies, but as a concept of power that is overdetermined by the ideology of everyday life. In this sense, they have made significant contributions to our understanding of the meaning – or emptiness – of race. There are divergent tendencies at play in these works. Some tend to ignore or insufficiently treat the scientific definition of race as a historical problem, or at least they do not delve very deeply into the *longue durée* of race. Other analyses are much more historical, but often present the concept of race solely in the context of the history of ideas. It is in these works that some argue that racialism is rational and 'at times' a useful tactic of those classified as racially inferior by the dominant ideology. Others suggest that racialism can only come from one source – one people – and nowhere else. Few if any of these works, even those histories of ideas, trace the concept of race along the entangled path that leads to the critique of science itself, choosing to reflect on philosophy and meaning. There is little attempt to continue the analysis through to its historical critique of truth. The scientific work is mentioned only in passing, but if reason and domination are connected, as they most certainly are, then the history of science must be of more than mere passing interest to the study of human variety. Reason – manifested through science and technological domination – is central to understanding race. This is especially true because the philosophy of race after Spencer came to serve as an interpretive adjunct to the science of race, just as philosophy came to be the adjunct of science. In the final analysis, the place to find the origins of the 'meaning' of race is in the sciences of life, in the magnificent bio-social discourse that spans disciplines from natural history to sociology. An investigation here leads one to understand the emptiness of the concept, except in relation to the formations of ideologies which serve as apparatuses for the deployment of scientific knowledge and its subsequent accumulation. In the current book, the analysis of race is not about finding the correct view of essential characteristics. This book is about how the social and biological sciences are invested with authority. In a more theoretical sense, it is an attempt to situate the study of race in the context of a more general study of bio-social discourse. There is no better expression of the ideological foundation of modern science than the history of the scientific classifications that form the parameters for most biological and sociological investigations of race and racial differences.

This book suggests an avenue of research that might fill a space made available by a variety of earlier work. An assumption that runs through much of this body of work is that the imperative to describe the experience of race can easily result in the distortion or elimination of the history of race. Relying on the description of this experience can make the history of race appear as a series of obvious and easily recognizable events that led naturally and inevitably to the present. If one must notice race, as some sociologists have claimed, then race must be seen not as a metaphysical concept or reduced to the level of mere identity. Instead, what we have come to refer to as race is a landscape of conflict. The extent of this conflict is limited by a bio-social discourse on the meaning of being human. This meaning is supposedly manifested by racial differences which we believe are constitutive of what kind of human we are, if indeed it allows that others are truly human. In the pages you are holding, race is examined not as a historical truth, but as a moment in the history of truth. This book is an investigation into the scientific classification of human variety. It is an attempt to recover a past which has been forgotten or repressed by the very sciences of life and society whose origins are to be found in these forgotten errors.

To examine human variety in terms of the history of truth requires that we understand the history of modern science and the history of race as interwoven histories. Does this mean that science is racist? Such a question rightly sounds absurd, but not for the reasons that the defenders of the privileges of scientific knowledge and the power of science would like us to believe, nor does the simplicity of the question negate its seriousness. To be sure a few scientists actively embrace racism, sexism, homophobia, etc., but the question is not whether science is racist. A more concrete question is: how does race function as a 'scientific ideology'? How has it successfully functioned as a scientific ideology for so long despite the considerable efforts that have been undertaken to make it a part of normal science? Race is a powerful expression of the attempt to fix the meaning of life and to determine its value. From their beginning, the human sciences have sought the fixed, unchanging meaning behind human diversity. Until Darwin, this meaning was not discussed in our terms of biological and cultural diversity because the naturalists did not think in those terms. Instead, what stood before them was not biology and culture but essentially one object of study which they understood as a history that was the playing out or unfolding of immanent and often racial determinations – *evolution* as understood in the *preformist* sense of the word. An acorn, as the old saying went, can only become an oak.

Historically, the language of race and the language of science reveal a continuity, even if the politics of this continuity are in constant flux. But *the ability of science to fix – however unstable and temporary this might be – the classification of human variety has contributed mightily to the establishment of the authority of the science of life in our understanding of the truth about human nature and society.* The authority of science to construct the degenerate, the criminal, the genius and the sexes as objects of knowledge is an authority intertwined with the scientific ideology of race and the administration of authority. This book is an investigation into race only in so far as race exists as a scientific ideology which is, in effect, a truth that is not quite true.[1]

One of many places to contribute to a broad project on the scientific study of human variety is the critical inquiry into the scientific classifications of human variety. One does this knowing that it is a question whose subject constantly refers it back to itself. One cannot escape creating classifications at the same time that one undertakes a critical study of fundamental systems of classification. Nevertheless, the task here is not to develop a theory of race, but ruthlessly to critique race as a scientific ideology. This makes the investigation of the history of classifications and their place in scientific ideology absolutely necessary to our obsession with finding the meaning of race.

> The history of science is the victim of a classification that simply it accepts, whereas the real problem is to discover why the classification exists, that is, to undertake a 'critical history of classifications'. To accept without criticism a division of knowledge into disciplines prior to the 'historical process' in which those disciplines develop is to succumb to an 'ideology'.[2]

It is not that race has priority over class, sex or gender, nor is it a more essential foundation to the classification of human variety. We might have found all manner of differences on which to base our production of human types and we have clearly used gender, class and other cultural differences. To argue for the priority of racial classifications would be to reinscribe the hierarchy that we should be attempting to make uninhabitable. A critical investigation of the many classifications of human variety discloses the history of scientific attempts to establish race. It is a history that should unsettle our most basic assumptions about both race and science. Certainly, nothing less than our faith in an immutable identity is called into question. At the same time, the manner in which science has described race and used it as a means to understand humans calls into question its own authority. 'The obsolete is condemned in the name of truth and objectivity. But what is now obsolete was once considered objectively true. Truth must submit itself to criticism and possible refutation or there is no science.'[3] The process by which we come to place a value on race rests within a hierarchical classification of bodies, attributes, truths and institutions. While classification is necessary for any production of knowledge, systems of classification also constrict and set the borders of acceptable knowledge. By investigating the scientific classification of human variety, we can begin to dismantle one of the ideological truths which we have since internalized and naturalized. This book is not concerned with denouncing the disciplines or reproaching them for their errors. Any discipline rests upon its particular regimes of truth: the 'report, naming, the narration of a Beginning, but also presentation, confirmation, explanation'.[4] Often, this includes how a genius lived, a discovery was made or a theory's predicted outcome was put to the test and resulted in a group of texts that established both an entire horizon of knowledge and the mythical history of the discipline itself. In contrast to this, the perspective that '[c]lassification is a condition of knowledge, not knowledge itself, and knowledge in turn dissolves classification'[5] neatly captures the process by which natural history and political economy became the disciplines of biology and society.

Smedley argued that 'the identification of race with a breeding line or stock of animals carries with it certain implications for how Europeans came to view human groups'. The very use of the term race placed an emphasis on innateness, on the unchanging and unalterable in humans. 'The term "race" made possible an easy analogy of inheritable and unchangeable features from breeding animals to human beings'.[6] The morphology and behaviour of humans defined and explained the animal as much, if not more, than the morphology and behaviour of the animal explained the human. Race now served as a foundation for creating a new creature: 'the European'. As a category for classifying humans within a general classificatory chart or table of nature, race has had varied and contradictory meanings. It was not the case that new principles of biology were applied to society, but that nature became social at the same time that the social became natural. The belief that they constituted each other took on new meaning after humans were placed in the world by Linné, Cuvier, Darwin and Marx. If the cause of human variety could be found, it would be the explanation for the variety of nature. It is also true that insights gained from animal husbandry could be used to explain human variety. These tendencies are common in the work of natural history that took 'the species question' as a central object of study.[7] Indeed, from natural history to biology, the quest to solve the species question was organized on the generally accepted assumption of multiple contemporary human species. The definition of species, a concept so central to the development of modern scientific thought, was determined within the context of the search for the origins of human variety.

It is not so much that race pervades everything, but that race is one mode of an all-pervasive bio-social discourse. Race is said to speak through the living being, who is subordinate to the truth of race. In so speaking, race could be said to pervade the social relations of everyday life: master/slave, creditor/debtor, capitalist/worker, parent/child, etc. But to naturalize race in such a way gives weight to an empty concept and distracts us from the apparatus of knowledge that speaks through and to race. This bio-social knowledge unites: 1. discourses on nature and life such as natural history, biology, medicine, ecology and their systems of classification; 2. discourses on the forces of life, which are of two kinds: first, the rational forces of Enlightenment – like the universals of history, consciousness and reason – and second, the irrational forces of the instincts, the mob, the anarchy of the social relations of capital, and the masses; 3. discourses on the stability of society, or social inertia, such as the sociological writings on stability, progress and degeneration. One might even hazard at this point to use again the word ideology.[8] Beneath the argument in this book is the assumption of a close relationship between authority and scientific ideologies. In one sense, the reader can take away from this work the most timid proposal, that scientific ideologies matter, that they have real effects in the world and that they are a part of the social relations they describe. A critical analysis that addresses scientific ideology would be impossible if ideology was not located in the materiality of everyday life, i.e., if it did not find expression in the materiality of social relations. What could express this more than the taxonomy of ourselves? What better than the classification of human variety that unites Camper's finding of beauty in the facial angle with your sideways glance at the suspicious or 'out of place' person walking through your neighbourhood?

Chapter 1 delves into the importance of the species question itself and the singular importance the riddle of human variety held in its investigation. The question of the existence of species and their origins would be decided by first solving the problem of human variety. It was believed that human variety held the key to understanding why variety existed in nature in general. We would finally know the reasons for our many differences in physiology, language and progress towards civilization. Monogenism and the fixity of species had an uneasy coexistence when variety was obvious to any observer and needed explanation.

Chapter 2 traces the shift from monogenism to polygenism, or the theory of multiple origins, i.e., the theory that each race originated at a different time and in a geographically isolated and unique locale. The monogenic theories were widely supported and derived most of this support from their seeming agreement with the Book of Genesis. Polygenic theories, on the other hand, were common amongst those who disputed the truthfulness of the biblical creation story and who were busy building the first respected scientific theory of human origins from the new world. The popular height of the polygenic theory was the publication of Josiah Nott and George Gliddon's *Types of Mankind*. This work was a tribute to their friend and teacher Samuel J. Morton as well as an open repudiation of religion in favour of free scientific investigation. Such investigation led them to proclaim the polygenic origins of human variety. This chapter gives a general discussion of the American School and an account of the brief period before Darwin when polygenism was the predominant scientific theory of the origins and meaning of human variety. This theory fell before Darwin's explanation for a common origin of humans and the chapter which follows puts Darwin in the context of the species question and his intervention against the monogenic/polygenic discourses. The title of Darwin's *Descent of Man* is itself an acknowledgment of the species question. It is suggested that when seen in the historical context of slavery and civil war, Darwin produced a sharp break not merely with polygenic theory, but with the entire discourse between the supporters of polygenic and monogenic theories. Darwin, it is argued, sought not only to produce a new scientific truth, but also to put an end to the current scientific discourse on human origins: 'when the principle of evolution is generally accepted, as it surely will be before long, the dispute between the monogenists and polygenists will die a silent and unobserved death'.[9]

Just as race does not pervade everything now, it should not be read back into our past. Race as we know it has not always existed, but it is no exaggeration to note that the dispute over human origins extends as far back in time as humans themselves, or at least what we know of our past. Not a few of us tend to see the past as rather homogenous, especially as we look back at those places where slavery was a social and economic foundation. As the historian of science Giorgio de Santillana pointed out, the ancients are all too often reduced to 'statues of Menon, uttering only one note'.[10] We know that this reduction is simplistic, although it has not prevented such simplicities from becoming the common-sense view. Just as there is no American view of human variety today, there was no one classical view of human variety. Because idealizations of the Greek polis are central to the concept of a Western tradition, it is worthwhile to explore various classical conceptions of human variety in its relation to slavery. This is not for antiquarian curiosity, but because the period when the American School's polygenic theory rose and fell was also the period when American slavery was defended and destroyed; in part by reference to the slave societies of classical antiquity.

Hesiod spoke of five races in his *Works and Days* which leads one to make a comparison to our current classificatory scheme of five races. However, Hesiod did not view race as an aspect of the body or of the diversity of humans at any one moment, but as a representation or even an expression of a historical era.[11] The key feature of Hesiod's view – his linkage of race to a specific historical period – did not survive its journey through the ages. In the centuries to come, it was replaced by a system of thought and governance which accepted race as a transhistorical essence. How this transformation occurred and why remains a subject of scholarly dispute, but it is not disputed that change happened. The racial series, too, was transformed from a fixed temporal sequence into an evolutionary scheme. And when this came about, the connection of race to history was recreated on the basis of race being an attribute of humans at a particular developmental stage of a progressive universal history. But how is it that Hesiod's view was transformed into a linear universal history on the one hand, and on the other an overdetermined cycle of history? The answer is that these transformations took place only when other objects and processes had been 'discovered' and could be read back into Hesiod's text. Hesiod's scheme of five races became a forerunner of our own classifications only when we confronted a new need to discover the forerunners of our own assumptions. Only when the taxonomists of the contemporary scientific era identified Hesiod as a significant but repressed predecessor, only then was the field of human classification marked on the one hand by the rise of scientific authority and on the other by slavery.

In Xenophon's *Oeconomicus*, human variety matters little except in so far as the differences between individuals could be functional for the efficient and just management of slaves. The successful estate would be one where the proper relationship between masters and slaves was consistently maintained. In his *Ways and Means*, Xenophon tells us that slaves are simply the best form of capital[12] because they are the most difficult form to steal. Unlike coinage they cannot be easily moved across borders if properly marked. 'How can you detect the export of public money? Money looks the same whether it is private property or belongs to the state. But how is a man to steal slaves when they are branded with the public mark and it is a penal offense to sell or export them?'[13] The 'brand of the slave' was a 'public mark' that distinguished the slave from the freeman or the foreigner. For Xenophon, the slave is not a slave by nature and is only marked by the branding scars on their skin. The slave cannot be identified by differences of skin, hair or bone, and obviously neither can the master. Master and the slave must instead learn to recognize in each other the social relations that dominate the economy of their lives.[14] This notion that a public mark is necessary to distinguish the master and the slave rests on the assumption that the origin of slavery is in the social relationship between master and slave.

It is Aristotle's view that has long been assumed to be the general view of Greek slavery. There are, he maintained, two fundamental forms of slavery: conventional and natural. Conventional slavery begins with the capture of the vanquished by the victors. The relation of the master and the slave is based upon circumstances of war rather than necessity. Conventional slavery implies no natural predisposition to being enslaved or to a 'slave mentality'. Conventional slavery allowed the enslavement of Greeks by other Greeks, and the 'naturally ruling' to enslave each other. So conventional slavery produces what Aristotle sees as an important problem of slavery: that at times the body of a slave looks like the body of a freeman, and vice versa. One can only imagine how the enslavement of whites in the American South by other whites would have been seen as equally problematic.

How can there be any need for a public mark designating one's social rank if nature itself produces such a mark? Since 'ruling and being ruled' is a social relation that is both necessary and expedient this 'distinction is already marked immediately at birth'. The Good are destined to be rulers – rulers of all other species, but certainly also rulers of others. It is either that or we are destined to be ruled by the better members of our species. If one doubts the necessity of this domination, then Aristotle suggests that one should first consider the relation of the soul to the body, and the body as being 'naturally ruled by the soul'. It might be true that there are individuals who are ruled by their bodies, or who are not up to the task of governing the actions of their body, but there are also 'corrupt' individuals whose pathologies can be measured against the nature of a master. Only those who are 'in a natural condition', i.e., the normal condition of the relation of the soul to the body, display the order 'that nature intends'. For Aristotle this is the point of reference from which all investigations must begin, 'for in those who are permanently bad, or who for the time being are in a bad condition, the reverse would often appear to be true – the body ruling the soul as a result of their evil and unnatural condition'.[15] The economy and morality of slavery is reproduced within the context of these natural and unnatural conditions. The lack of a public mark does not appear to prevent the Aristotelian observer from distinguishing between the naturally ruled and the natural ruler. One does not have to observe their everyday lives to make judgements about their value. What holds true in one's inner life is manifested to the outward world, the bad soul is marked by the bad body. This distinction is not one chosen nor can it be given except by nature itself. To be ruled and to rule is the nature of the relation of the body to the soul, of the animal to the human, of the woman to the man and the child to the parent. This relation is not only necessary but practical and expedient for the arrangement of life. The tame animal cannot survive without the stewardship of its ruler providing the 'benefit of preservation'. So too is the relationship of male to female. If we agree with Aristotle, the ubiquitous relation of master and slave becomes a general fact of social life. Just as the tame animal owes its preservation to its master – and its master displays mastery through the ability to preserve its inferiority – the slave is better off being ruled precisely because the slave is destined by nature to be ruled. No master is capable of enduring the position of slave, and no slave can function as a master without the unnatural condition of the relation manifesting itself as injustice. What holds true in terms of mental differences also holds true in terms of obvious physical differences. The relation of the soul to the body and qualities of mental superiority and inferiority are manifested by the shape and condition of the body. '[T]he ruler is always a freeman' since the condition of being a slave, i.e., to have allowed oneself to be enslaved, is impossible for the freeman to endure. Nature produces a physical difference between the body of the freeman and that of the slave – 'giving the latter strength for the menial duties of life, but making the former upright in carriage and (though useless for physical labour) useful for the various purposes of civic life – a life which tends, as it develops, to be divided into military service and the occupations of peace'. There is a principle that runs through each of the fundamental relations of existence – soul and body, human and nature, master and slave, man and woman, parent and child, etc. – that 'what holds good in man's inner life also holds good outside of it'.[16] If we can distinguish beings by the quality of their natural differences, we might use the healthy as the standard by which we measure of all others. The healthy person reproduces the healthy relation between body and soul, and between heaven and earth. This quality is their mark. It is not a 'public mark' bestowed by humans, but a mark endowed by nature. The healthy being stands in sharp relief to the teeming background of 'those who are permanently bad' and those 'who for the time being are in a bad condition'. It is in these two classes of humans in whom we 'find the body ruling the soul as the result of their evil and unnatural condition'. If goodness is defined by and defines the typology of the body, then classifications can be produced which are constitutive of scientific ideologies. It is no exaggeration to maintain that Aristotle and nineteenth-century science share a common logic regarding nature, domination, the body and goodness.[17] The soul differs from the body, the body of the human differs from the animal, the male from the female, the adult from the child. Those endowed with the ability to rule must do so else the very structure of nature will be unhinged. The relation of master and slave is the condition of ruling and being ruled which is not only necessary and expedient, but its logical result is the political assertion that domination, classification and subordination are present at the very origins of Western democracy. Is it really any surprise that Aristotle's vision of uniting the Greeks of good souls and bodies resulted in the empire of Alexander?[18] Uniting Aristotle and Xenophon was a recognizably modern sensibility where races of man are defined by their double relation to the state and to nature. Now questions such as 'What type of human is this?' 'How does it live?' held the keys to understanding society. Above all, the best means to understand these relationships was found in the creation of a proper system of classification of visible characteristics. The questioning of the meaning of humans and the will to construct a taxonomic system made it possible for natural history to claim Aristotle as its own, as it did in the eighteenth century. There is a profound convergence between Aristotle's view of the innateness of slavery and Hegel's comments on Africa and its Negroes in his *Philosophy of History*.[19] While Hegel rejected the conflation of inner and outer that he saw in physiology and phrenology, he too found that the naturalness of slavery marked the slave. By that point in time, this essence no longer marked the general category of the barbarian, but the more specific and naturalized category of the African, whose natural attributes are said by Hegel to make them particularly well suited to slavery.

Aristotle's view reaches its historic end with Hegel's *Philosophy of Right*:

> Hence in this identity of the universal will with the particular will, right and duty coalesce, and by being in the ethical order a man has rights in so far as he has duties, and duties in so far as he has rights ... A slave has no duties; only a free man has them.[20]

The translator T. M. Knox makes no comment on this section, but we might interpose that if Hegel is the final culmination of the Aristotelian view of the innateness of slavery, then we have discerned a fundamental continuity between the ancients and ourselves. To any who fancy themselves anti-racist, the history of racialist ideologies and racism could be neatly contained in a 'Western tradition' culminating in modern science and the scientific understanding of human nature. But the story is more complicated than that, for the ancients were no more monolithic than we are; their selection and appropriation by later generations is always determined by the needs of the moment. Rome, we often forget, was the capital of an empire stretching from England to Persia and encircling the entire Mediterranean from the Danube to the Sahara and from the Pillars of Hercules to Palestine. Moreover, Rome extended the franchise to all its people. Every imperial subject could potentially become a citizen:

> Rome gradually became the common temple of her subjects, and the freedom of the city was bestowed on all the gods of mankind. The narrow policy of preserving, without any foreign mixture, the pure blood of the ancient civilizations, had checked the fortune and hastened the ruin of Athens and Sparta. The aspiring genius of Rome sacrificed vanity to ambition, and deemed it more prudent, as well as honorable, to adopt virtue and merit for her own wheresoever they were found, among slaves, or strangers, enemies, or barbarians.[21]

This diversity and openness was blamed for its collapse by those nineteenth-century historians whose first and last concerns were with the potential fates of their own empires. The fact of their asking the question was just as important – in terms of its political meaning – as their answers. The desire to understand or read their destiny through the history of earlier empires is easily understandable, but why did Rome figure so prominently in their discussions? This is not as simple a question to answer as it might seem. Certainly the education that children of the upper and middle classes received deserves some scrutiny here. For them, classical training was the norm both in schools and in the military academies. The older conflict between the ancients and the moderns was not wholly resolved in favour of the moderns regardless of the overthrowing of the authority of the ancient texts to establish the parameters of knowledge. With the rise of natural history, Aristotle was no longer the unique guide to nature. But having been overthrown, ancient knowledge was restored to a new position of importance, not as the final authority, but as a marker of our progress. It was overcome, but still served as the foundation for all that we have today. The diversity of the Roman empire seemed to mirror the diversity of the modern European empires and the degeneration of European society from its contact with its colonies was an echo of the decay of the imperial order.

> What lay behind and constantly reacted upon Rome's disintegration was, after all, to a considerable extent, the fact that the people who built Rome had given way to a different race. The lack of energy and enterprise, the failure of foresight and common sense, the weakening of moral and political stamina, all were concomitant with the gradual diminution of the stock which, during the earlier days, had displayed these qualities.[22]

The need to explain and understand human diversity was not lost on the Romans. It was a topic in everyday life and elite discourse.[23] Tacitus certainly goes about his description of the Germans with an eye towards explaining not only the consequences of Rome's moral corruption, but also the difference between Romans and barbarians in such a way as to elevate and extol the Roman virtues which he found to be on the wane. But others, most notably the Elder Pliny, whose encyclopedic *Natural History* is an collection of all the knowledge available to him, shared neither the concerns nor the asceticism of Tacitus.[24] In contrast, Pliny had 'a Roman tolerance for and joy in human diversity', an attitude that is clearly present in Book VII, where he describes the so-called 'monstrous races of men'.[25]

> And indeed, I gave the whole range of races of humans in my telling of the kinds of nations. And I will not at this time give examples of the rites and morals which are too numerous to manage [in this work] and are nearly as many as the groupings of humans ... Truly, Nature, powerful and majestic in its momentous whole lacks real weight if one considers only its parts and not the complex spirit of the totality. Our undulating fear of tigers, panthers, peacocks, of the spots and colorings of so many liv-

ing creatures calls to mind a minor truism but one of immense value: so many types of speaking, so many languages, so great are the varieties of eloquence that 'a barbarian can not pass through here in the place of a civilized person!' Now unto the face and the features there are in common only ten or perhaps a few more parts, but there are no two likenesses existing in the so many thousands of humans that can not be distinguished – something that no art might construct in such an abundance and from so small a number of parts![26]

Plinian races were not just beings of faraway lands such as India or Ethiopia – which were sometimes the same thing in the ancient texts – but could be found near at hand in Northern Europe, Libya and even Italy. Just as Pliny expanded the varieties well beyond the ten or so mentioned by his sources, so too did the medieval commentators, cartographers and scribes proliferate the number of Pliny's prodigious races and they sometimes conflated separate races or abbreviated the description of their distinguishing characteristics so as to find two types where Pliny had identified only one.

Despite the influence of Pliny's prodigious types, the great gulf which separates Pliny from modern natural history exists because his catalogue of types is not a system of classification. Pliny offers an exhaustive review of the sources available to him, but he consistently leaves open the possibility of new discoveries and new interpretations. The Plinian races formed a rich field of possibilities that exceeded the limits of knowledge itself. The various attributes of humans could be combined to produce a seemingly infinite variety of humans. They could be multiplied into the various monstrous races through language and physical attributes. The differences would not only be obviously physical, 'those of hair, bone, and skin', but also of custom, so that 'those who stood staring at the sun' or 'those who eat human flesh' constitute a separate 'race' of human in the same way that a Saropod or a Cyclops do. For Pliny, the diversity of races and the similarity of the individuals is caused as much by chance as by parentage.[27] He was not interested in the regular, or at least had neither the time nor space to enumerate the contents of the world. In several passages Pliny says that the regular or usual is obvious and need not be described in detail precisely because it is the same everywhere and a part of everyone's experience: it is the same 'over there' as it is 'here'. It was the unusual, the rare, the prodigious to which he devoted his attention. The relative stability brought by the power of the empire during Pliny's time would have perhaps made human variety less threatening, especially when we consider the size of the empire and even the city of Rome itself relative to the Greek polis. Trade routes were extensive and communication throughout the empire relatively secure and efficient. Although both were dependent on slavery, the openness of Roman society was noticeably lacking in Greece.[28]

Certainly Pliny's fellow Stoic, Seneca, shared this openness. His Letter XLVII, also known as *Treating Slaves as Equals*, speaks in much the same voice.

The differences amongst people are those trappings of clothing and social class, but all are the result of fate because we are all subject to fate; masters are themselves slaves to 'monstrous greed', ambition and the workings of fate. Seneca did not oppose slavery as an economic institution, but says that 'Each man has a character of his own choosing; it is chance or fate that decides his choice of job' and that everyone should be judged 'according to their character, and not their jobs'. Thus he can write to his 'dear Lucilius':

> Slaves! No, they are our fellow-slaves, if one reflects that Fortune has equal power over slaves and free men alike ... 'He is a slave'. His soul, however, may be that of a freeman. 'He is a slave.' But shall that stand in his way? Show me a man who is not a slave; one is a slave to sex, another to greed, another to ambition, and all men are slaves to hope or fear.[29]

Pliny's *Natural History* and later natural histories share the same structure in their presentation. In a work such as Pliny's, language is not secondary or an afterthought attached to the description of the body, but is part of the body itself. It would remain so for a long time to come, through the time of curio cabinets and Linné's fixing of nomenclature within the classification of nature.[30] The inability to limit nature made classification as it had been carefully laid out from Linné to Agassiz an impossible task. It is important then to consider this very transition from infinite variety to a natural history characterized first by description and second by the fixity of language and classification. This was not a smooth transition: the species question and classification were ways of coming to terms with the problem of variety. This problem of variety, in particular human variety, stands at the centre of the species question and generates schemes of classification which we still use today to capture the complexity of nature.

The question of continuity and discontinuity is a problem that the histories of the scientific ideologies of human variety can throw a great deal of light upon, and critically situate our contemporary scientific ideologies of race. So, in what follows, the problem of continuity and discontinuity is to be found in the use of concepts such as tradition, community, influence, development or essence. The use of these keywords remains an obvious characteristic of scientific ideologies of race, and so it follows that these concepts cannot really aid in understanding the changing meanings of human variety. At the same time, the notion of a break is too easy to assume and allows a certain notion of fixed periodizations in through the back door. The notion of a break makes it possible to repress from memory and disciplinary histories all the former truths and 'wretched knowledges' that once defined Truth. A break in one field does not mean that there is a simultaneous break in or across all other fields. Some fields may have no ready or direct relation to each other, or the break is posthumously located in the work of an author: one could have Darwin *or* Wallace; or communism without Marx.

Mendel's paper sat unread for almost fifty years and was only discovered as the foundation of genetics after the formation of the field.[31]

In this study, we are dealing with myriad breaks and discontinuities that occur in the midst of the enormous continuities of knowledge production, with recoveries of lost knowledge and with the appropriation, repression or destruction of 'wretched subjects'.[32]

The classical view took for granted the variety of nature and assumed this variety to be a fundamental aspect of a world that, at least for the Epicureans and Stoics, was infinite. With the end of antiquity, this variety was seen as a relation solely between nature and its creator, and with it, human variety became symbolic of the divine will and judgement. Finally, this continuity was ruptured in the nineteenth century by the appearance of the enlightened man restored to his place in the natural order (Darwin) or as a prodigious type, or as the totality of the Divine (Agassiz). This all happened, of course, in relation to the emergence of that new object of knowledge: life. But these general statements on continuity are not particularly insightful or original, for it is a commonplace of our time that the question of chronology receives much serious analysis. Nor can these statements tell the story of how human variety came to be understood according to racial types. Pliny's 'prodigious' humans are humans. They are not deviations, degenerates or deformities, nor are they the result of hybridization. They are not signs or omens, representations or divine favour or wrath. This is a fundamentally different conception of human variety from the Christian one. Monsters marked the limit of nature, but prodigious varieties did not. What Marx found in capital Pliny found in nature: that the process of accumulation and reproduction marks its own limit.

In Aristotle's works human variety was not a defining characteristic of a particular historical period. The ambition was to define and classify humans through the doubled relation between polis and nature, and between humans and nature. It does not minimize this double relation to recognize that the social relation of master and slave is seen as fixed and permanent and allows race to take on a transhistorical presence, for a slave is now born to be a slave, and the master a master. The everyday social relation between humans in a polis transcends its material basis and comes to stand for all human relationships. On this point Marx and Nietzsche agree: a fundamental social relationship is that of creditor/debtor, and this relationship has a history that could be excavated through critical and genealogical approaches. The basic social bond that we bring to light is all too often one of cruelty and cooperation – because breaking or transvaluing the already given social relation reproduces the same cruelty that brought it into existence in the first place.[33] The return of the repressed appears in its most concrete form: the cruelty of the relation never disappears, it simply comes to appear as nature itself, or as the very definition of what is natural in terms of human nature. Nature comes to embody the cruelty found in the State and in nature and vice versa.

> The welding of a hitherto unchecked and shapeless populace into a firm form was not only instituted by an act of violence but also carried to its conclusion by nothing but acts of violence – that the oldest 'state' thus appeared as a fearful tyranny, as an oppressive and remorseless machine, and went on working until this raw material of people and semi-animals was at last not only thoroughly kneaded and pliant, but also *formed*.[34]

The relationship of slavery to our understanding of human variety is just one specific instance of the broader apparatus of cruelty and cooperation.[35]

The complexity of its subject makes a work such as this one difficult and its conclusions ultimately tentative until such time as others take up the task. There are some similarities and alliances that one might expect, but there are also many that are surprising or ironic. What we can definitely say is that this complex arrangement of discourses and institutions producing – and produced by – the knowledge of human variety teems with continuities, discontinuities, dialectics, fanciful speculations, frauds, empirical observations, measurements and classifications. At the very least we can conclude that out of this emerged the authority of the sciences of life and society: biology and sociology as the true sciences of Enlightenment. The assumption underlying this work is that the true meaning of human variety and its origins are like the Ghosts of Africa mentioned by Pliny at the end of his catalogue of the notable varieties. He describes them as the species of human that vanish when approached.

1 CLASSIFICATION AND
THE SPECIES QUESTION

The Species of Man: Physiology, Language and Civilization

One expression of colonialism and imperialism was the imposition of a nomenclature and taxonomy on the newly discovered flora and fauna. This new order was no longer depicted by medieval stonemasons, gardeners and scribes, but by cartographers, explorers, travel writers, slavers and those ultimately burdened with the duty to spread enlightenment. The discovery of the New World produced a great rupture in the understanding of nature, and the European expansion brought within its sphere new plants and animals which vastly expanded the known variety of nature. If the world was already too varied for Pliny to catalogue, what was to be done now that an entire New World had to be incorporated into the map of creation? To produce a system whereby different naturalists could refer to individual examples of plant and animal varieties and be mutually understood as referring to the same object became a fundamental scientific quest. With the necessity for a standard system for naming and classification, 'the systematic classification of the natural world emerged as one of the quintessential achievements of modern science';[1] and it was of course Linné's great achievement. At the very moment when our domination of nature could be cheered alongside it, 'the paradox between the discourse of freedom and the practice of slavery marked the ascendency of a succession of Western nations within the early modern global economy'.[2]

The histories of the life sciences all linger for considerable time over the topic of classification. It is often noted that Linné's classification allowed for a scientific description of the natural world. It is also noted that the acceptance of the Linnean classification marked the acceptance of man as a part of the natural order. The recent attention to Linné's work stresses the social and local political objectives.[3] In the need to classify the specimens of exotic animals and plants was manifested also the parallel need to classify culture. Linné's table presented in outline the knowledge of human nature at the moment the sciences of life and society appeared on the horizon. Both natural history and the sciences of life gained – in their own times – the status of science because they claimed to explain difference in nature. However, natural history failed in many respects to provide an adequate explanation of difference and of the species question. It did succeed in setting the variety of man as the key for explaining the variety of nature by placing man in nature. The classification of racial types and the classification of language by historical/comparative philology should be seen in the context of the classification of specimens, and of culture as civilization.[4]

The order of nature would not emerge from the everyday human experience of it but through the rational analysis of its internal structure. It was assumed that this internal structure is available to human understanding and technological domination precisely because nature and mind recapitulated each other. Ontogeny recapitulated phylogeny and so by necessity the structure of nature had to be fixed and encapsulated in a table of classification. From that moment on, the nomenclature captured nature in the rationality of language. The history of language and the history of nature developed a complex mutualism: the perfection or degeneration of one would determine the limits of the other. Along with this transformation, the history of civilization became inseparable from the history of human variety. The naturalists of the time knew this and pursued their work in philology, taxonomy and ethnology with the general expectation that advances in one domain would benefit the others. The production of knowledge was unified in the quest for the origin, development and meaning of man. By this route the desire to establish the meaning and origin of human variety gave and received meaning from the linkage of race, language and civilization. The discourse on the Indo-European or Aryan 'invasions' seemed to establish the origins of the European either in Europe or in nomadic horsemen from the Steppes. Either origin could serve well enough as a basis for political dominance. The invaders from the North – or the East – were now to be remembered in texts and songs composed long after and penned by those who saw themselves as descendants of the aboriginal Indo-Europeans themselves. The work of the Grimm brothers and other linguists like Rydberg and Rask helped establish the legitimacy of the Teutonic Origins theory for use by a rising Germanic nation-state; and this collaboration reached its apex with the ambiguous work of anthropologists such as Rudolf Poch and the explicit work of the later Indo-Europeanists of the Ahbeneber to establish the final solution to the species question.[5] It was to be the most highly developed species of man, in the most advanced civilization, with the most efficient means of communication and terrifying military that would finally propose the solution to the question of the varieties of man.

From the sixteenth to the eighteenth centuries, 'race' developed as a classificatory term similar to – and at times interchangeable with – 'people', 'nation', 'kind', 'variety', 'stock' and so forth. But these words did not always mean the same thing or refer to the same phenomena, nor were they always so easily interchangeable. By the last half of the eighteenth century, investigations, clas-sifications and definitions of human populations coincided with the elevation of 'race' to a status where it could become a means for a scientific classification of humans. Like the machine gun, race was first deployed against non-European groups and soon after against those Europeans who varied in some way from the norms of bourgeois life.[6]

Simply titling a work *The Origin of Species* was enough to stake out a theo-retical position and to situate the work within several mid-nineteenth-century conflicts. Two of these are of immediate interest: the species question, i.e., the *meaning* of variety or difference as we encounter it today; and the problem of the origin of species, i.e., the *history* of variety. At times, either one might have been taken as the central problem, but the relationship between the two was always quite fluid. Both discourses had to be engaged in order to understand the con-temporary views of human variety, and if race seemed to establish a coherence to the history of nature, it was because race could be said to have existed from the first distant moment of origin, and yet it – race and its origin – still appears everywhere and at every moment. Race is so reified and alienated that it appears to us as an animating force which lives through us, as the truth of the gods was once said to animate the poet's voice. This fetishism is the essence of the scientific ideology of race: it is ahistorical and yet is said to be a force driving our history.

If race is ever present and 'must be noticed', and if it is really reproduced in the variety of human physiques, then the horrors of the origin of race are ever with us and reproduced by us.[7] The slavers of old would be proved correct, and we would find ourselves left with nothing better than responsible eugenics and enlightened domination, a utopia still dear to many of our own day. The origin of the group would be recapitulated in the origin of the individual and any break-down or confusion of the distinction between the individual and the group would be a detriment to both. There would be little rhetorical difficulty in elevating this empty belief into notions of ancestral heritage and doctrines of blood and soil.

The dogma that the natural repulsion felt for the other was the foundation of group or national solidarity can only be maintained if the attributes of species never change. In the encounters of the European with the many others, the ques-tion of origin was not so much about the origin of the other as about the origin of the superior and morally mature Europeans. Evidence of universal history was easily at hand in the realm of politics and economics. In science and education the development of children and primitives was understood to recapitulate the stages of the European's development, while the woman and the deviant came to represent the pathological failure to develop, along with the corrupting, degen-erative, atavistic forces still within modern society.

Establishing the origin of the European was to find the key to its fate, and so it is not surprising that the Indo-European problem became so closely allied with the species question. Both questions produced discourses that cut across disciplines, undermining some and founding new ones which further interpenetrated each other. 'From whence had the European arisen?' was not a simple or simple-minded question. Moreover, the answer, after all, had to justify an entire history of the world whose beginning and end was nothing less than the production of European man. The Indo-European problem was not simply one of locating the geographical origins of those languages and peoples classified as Indo-European. The early anatomical and philological evidence appeared to point to multiple origins. A geographical region stretching from the Caucuses and the Steppes represented the continuity of nature as well as a revolutionary break in human history. Europeans were seen as having a common origin in the nomadic Aryan – nomadic, it was assumed, because of the number of words referring to horses and nomadic life, and almost none referring to pastoral living in the reconstructed Indo-European language. Though it receives more attention, those arguing for the emergence of the Aryan in northern Europe or central Asia all agreed that they were talking about the same variety of human, and that this variety had an origin that gave it a unique heritage linking language and physical type. The debate that was joined between those advocating European or central Asian origins was not a debate about the uniqueness or the existence of the Aryan, or over its nomadic lifestyle, but over the geographic origins of a unique form of human. The resolution was for many to be found in physiology, philology and even in the land and sky: 'In taking a grand view of the subject, a mystic harmony was found [by Jakob Grimm] to exist between the apparent course of the sun and the real migrations of peoples'.[8] The origins of the Aryan could be found in the 'linguistic movements' and the human varieties which give us a map of natural history:

> the movements of the Teutonic races were from North to South, and they migrated both eastward and westward. Both prehistoric and historic facts thus tend to establish the theory that the Aryan domain of Europe, within definable limits, comprised the central and northern part of Europe ... Farthest to the North ... must have dwelt those people who afterwards produced the Teutonic tongue

Only two pages later this linguistic movement materializes in physiques and psyches:

> The northern position of the ancient Teutons necessarily had the effect that they, better than all other Aryan people, preserve the original race-type, as they were less exposed to mixing with non-Aryan elements ... The Teutonic type, which doubtless also was the Aryan in general before much spreading and consequent mixing with other races had taken place, has, as already indicated, been described in the following manner: Tall, white skin, blue eyes, fair hair. Anthropological science has given them one more mark – they are dolicocephalous.[9]

Locating Germany as the ancestral home of the 'original race-type' did not nec-essarily rule out that idea that the type originated elsewhere. The geographic range of the Indo-Germanic or Indo-European languages left open the possibil-ity that the nomadic ancestral Europeans were immigrants. This was the core of that other expression of the species question, the Indo-European problem. If Aryans were not indigenous to Europe, where were they from? Whether indig-enous or not, the superiority of the Indo-European could be easily demonstrated by the European nation's domination of much of the globe. The search for the proto-Indo-European homeland would explain how Europe became the domain of the European, a domain it constituted as it constructed itself. The origins of the European and the origins of civilization became the same search, one 'central to any explanation of how Europe became European. In a larger sense the search for the origins of European civilization.'[10] Michael Wood's documentary on the Ahbeneber and its expeditions to South America and Tibet – and later human experiments in the camps – is instructive.[11] The origin of the Aryan had to be unique and unrelated – except by strife and conquest – to the origins and histo-ries of the other races.

The relationship of languages to each other, the course of the sun from east to west, and prehistoric nomadic migrations explained how the Indo-European languages derived from the language of the Aryans. For the brothers Grimm, this meant that language had degenerated from its pure state in much the same manner that bodies and civilizations had degenerated. Now language could have its own natural history. The Indo-European family of languages constituted the artefacts of the history of civilization. 'The reader might thus be led to know the origin of most nations, by tracing words to different places, and thereby find out the source of their languages.'[12] William Jones, who showed that San-skrit belonged to the Indo-European linguistic family and was perhaps a close descendant of the original language of the Aryans, put it in these terms:

> The Sanskrit language, whatever be its antiquity, is of a wonderful structure; more perfect than the Greek, more copious than the Latin, and more exquisitely refined than either, yet bearing to both of them a stronger affinity, both in the roots of verbs and in the forms of grammar, than could possibly have been produced by accident; so strong, indeed, that no philologist could examine them all three, without believ-ing them to have sprung from some common sense, which, perhaps, no longer exists: there is a similar reason, though not quite so forcible, for supposing that both the Gothick and the Celtick, though blended with a different idiom, had the same origin with the Sanskrit; and the old Persian might be added to the same family.[13]

The use of a genealogical family as it developed in philology had two features worth mentioning here. First, the use of the concept of linguistic families carried with it an assumption of change or modification, i.e., descent and genealogical association. Second, the genealogical family already contains the assumption of development and extinction. As with peoples, civilizations, plants and animals, so languages live and die. 'There is really only one 'speaker who never ceases to speak' and that speaker is always 'the people'.[14] The history of the civilization is the life of its people, and a language and a civilization are the life's work of a people. Thus '[e]very usage possessed its importance as an expression of the folk-spirit'.[15] So too was the spirit of civilization understood as the accumulation, either positive or degenerative, of the labour of its people. This essential connection between them allowed each to explain the other.

> As any language has been in some way connected with a preceding one so may every species. Languages may improve & may degenerate, may change with comparative rapidity, or be persistent for indefinite periods – & once extinct can never reappear. All this may be true of species. It may happen in the course of ages that as some words & forms of speech & grammar are always growing obsolete & new ones invented, the languages coming from one original stock may have nothing in common to the parent tongue from which they branched off & yet if they could be followed up historically the bond would be discoverable. Unity & Continuity, therefore does not of necessity imply a discoverable relationship in every true existing language & it may be so perhaps in regard to some living species, which at times indefinitely remote, sprang from a common progenitor.[16]

Some writers could not ignore the implications of taxonomic work for the search for the homeland of the Aryans. In 1851, R. G. Latham turned from the study of the classification of races to the study of the Indo-European languages. Latham argued that organizing languages and dialects in terms of genus and species allowed one 'to derive the Indo-Europeans of Europe from the Indo-Europeans of Asia [which] is the same thing in ethnology as if in herpetology one were to derive the reptiles of Great Britain from those of Ireland'.[17] They both derived from a common nomadic source, that is why the analogy moved from language to biology and back again to language – and not as we might expect from biology to language. The genealogy of language served as the model for the genealogical description of species. Much more than a romantic *esprit* of the times, this powerful conception of genealogy explains how someone like Latham could move so easily from racial classification to philology.

The publication of Adolphe Pictet's *Les Origines indo-européennes* marks the beginning of 'a more scientific search' for Indo-European origins. It was an obvious – or perhaps deliberate – understatement on the part of a later commentator that this work was 'not without consequence for later research'.[18] Pictet's work referred to the ancient Aryans as 'privilegie entre toutes les autres par la beaute du sang, et par les dons de l'intelligence'.[19] Pictet's argument for Aryan superiority was only one of a great many works that added to the rapidly accumulating discourse on the racial superiority of the Aryan. Like so many other works, it emerged during a long campaign to establish the origins of Europe in terms of race, language and culture. It should come as no real surprise today that those who found the Aryan to be superior also tended to locate the origins of the Indo-European in Northern Europe. 'All the tribes of the earth belong to the Negro or the Mongolian race ... the Caucasian is pre-eminently the man of civilization.'[20] In the nineteenth century, philology had powerful political and social ramifications.[21] Pictet proposed to develop a lexicon of proto-Indo-European that would allow us to understand the world of the Aryans as preserved in their vocabulary. This proto-lexicon, it was thought, would be a means to locate the *urheimat* or homeland of the Indo-European. Pictet termed his method of studying the residues and traces of extinct language 'linguistic paleontology'[22] to emphasize the continuity between linguistic and anthropological work.

Jacob Grimm declared the common origin of 'Poetry and Law' in an early essay by that title. They had, he said, 'grown in one bed'. The poet and the judge uttered the very thought of the German people.[23] The association of the name Grimm with a discussion such as this probably surprises many casual readers today. For many, the Grimms bring to mind cherished memories of childhood, of fantastic stories read at bedtime, of fables of morality and the almost inevitable triumph of good over evil. There is no doubt that morality and childhood are at the centre of the work of the brothers Grimm, but if we rely on the sanitized rendering of their collections, we will never approach the true significance of childhood, morality and history to the scholars of their day. The childhood that they sought and invoked was the childhood of the Aryans, whose nobility and superior morality had become corrupt and degenerate. From law to poetry to myth, language itself displayed this degeneration. All that the modern Germans called their culture was little more than corrupted artefacts. The contemporary German had to be understood as living amongst the ruins of a society and within the ruined bodies of a once noble *volk*. All was not lost, though, and what remained could be preserved and through this work of remembrance the people might be restored to their past vigour. This broad social revolution could be, and in fact had to be, reconstructed with the same methods as one had constructed the language of the Aryans, i.e., by moving backwards towards its pure origins and glorious past. 'Every passage possess its importance as an expression of the people.'[24] Law, language, poetry, myth and the folk spirit had all degenerated along similar lines. While Grimm's Law provides a genealogical description of a line of modification, it is really descent with morbid results – a genealogical record of linguistic and social decay and degeneration.

The historical analysis of the Indo-European languages demonstrated that just as

> in language, the further we are able to follow it, a higher perfection of form ... a like proceeding must be justifiable in mythology, too, and from its dry water courses we may guess the copious spring, from its stagnant swamps the ancient river. Nations hold fast by prescription, unless we spread under it a bed of still heathen soil.[25]

There exists an unbroken line between Grimm's work and that of others who hoped to solve the Indo-European problem by establishing the Aryan homeland in Northern Europe, for

> [w]hen historical facts to the contrary are wanting, then the root of a great family of languages should naturally be looked for in the ground which supports the trunk and is shaded by the crown, and not underneath the ends of the farthest reaching branches ... the great mass of Aryans live in Europe, and have lived there as far back as history sheds a ray of light. Why then, not apply to the Aryans and to Europe the same conclusions as hold good in the case of the Mongolians of Asia? And why not apply to ethnology the same principles as are admitted unchallenged in regard to the geography of plants and animals? Do we not in botany and zoology seek the original home and centre of a species where it shows its greatest vitality, the greatest power of multiplying and producing varieties? These questions, asked by Latham, remained for some time unanswered[26]

The question was answered in the affirmative by Latham, Rydberg and the Grimms. For the brothers Grimm folklore was the cultural detritus of ancient Germanic belief. The myths of ancient times had disintegrated first into heroic legends and romances and then into folk tales. The degenerate remains preserved an ancestral heritage they wanted to recover as the basis for a new Germany. 'All my works relate to the Fatherland, from whose soil they derive their strength.'[27] Language united a people with their civilization; it gave meaning to their history and recorded the past. This recording is not only found in texts and documents, but also in the history of the sounds of languages, in the phenomenon of linguistic change described by Rask, Grimm and Verner.

> With Grimm the stress lies decidedly on the inner reason connecting the various parts of the shifting. He felt able to set forth a single law incorporating all its phases ... The cause and time of the Consonant Shift are moot questions but it is probable that they bear a relation to each other. It is hardly by mere accident that this strikingly comprehensive and homogenous group of phonic changes is contemporaneous with what may justly be called the most momentous national movement in history: the Germanic Migrations ('Volkerwanderung')[28]

Grimm sought the cause for this shift in the 'impetuous character' of the ancient German tribes. Prokosch did not follow Grimm in succumbing to the 'worst fallacies of the geographical and ethnological theories' of his day. Psychological or racial origins are not source of the consonant shift as its 'social, economic, and emotional background'. Nevertheless, he agreed with Grimm that 'the Consonant Shift appears to stand in chronological and causal relation to the ... Volkerwanderung'.[29]

Rydberg's mention of Latham's work is not merely a passing reference. Latham's belief in the European origin of the Aryans is approvingly contrasted with Pictet's and Max Muller's view of the Asiatic origin. Rydberg's own work rested on the general assumption of monogenism, but he also insisted that 'there has been *a European-Aryan country*. And the question as to where it was located is of the most vital importance, as it is closely connected with the question of the *original home of the Teutons*' (original emphasis). Linguistic evidence favoured the European origin of the Aryan, Rydberg said, but it was not yet conclusive. Although the Asiatic theory continued to be more generally accepted, the proponents of the European origin had a least succeeded in raising significant concerns that forced the proponents of the Asiatic origin to modify their theory significantly to grant that while the origin of the Aryan was in Asia, the culture might have over time become uniquely Germanic. Instead of migrating from Asia in linguistically distinct groups, the Aryans were now taken to have 'formed one homogenous mass which gradually on our continent divided itself into Celts, Teutons, Slavs, and Greco-Italians'. For Rydberg, the origin of the Teutons 'in the Central and Northern part of Europe' was not inconsistent with the Asiatic origin of the Indo-European.[30]

Rydberg went beyond the linguistic evidence in his efforts to establish the uniqueness of Teutonic mythology. He took as his evidence work in craniology and anthropology in order to bolster his argument. In what Ibn Khaldun called the Northern Wastes, the ancient Teutons 'were less exposed to mixing with non-Aryan elements' and so were able to maintain the purity of their 'original racial type'. Ancient histories described them as 'Tall, white skin, blue eyes, fair hair' and anthropology gave them 'one more mark – they are dolicocephalous'. The most 'pure types' exhibiting this characteristic were 'the Scandinavians, Northern Germans, some English and Dutch'. Relying on the craniological measurements of Welcker and studies of the distribution of blond and brunette types, Rydberg employed what he saw as Latham's 'strict methodology' to mark out what he believed to be the unique identity of the Teutons by shedding 'light on the beliefs and ideas that existed in the minds of our ancestors ... to distinguish between older and younger elements of Teutonic mythology, and to secure a basis for studying its development through centuries that have left us no literary monuments'.[31]

Rask also showed an interest in uniting philology and natural history. He was 'acquainted with natural philosophy, specifically the classificatory system of Cuvier' and there are 'many striking similarities between Cuvier's and Rask's typological concepts'. It became Rask's goal to seek the historical developments 'which permitted the typological classification of languages based upon their grammatical structures and cogent and rigorous formulations of rules for determining linguistic relationships'. The basis for this classification was to become better known as Grimm's law. Rask was the first to seek the underlying framework of grammatical rules to which all linguistic behaviour must conform and to 'perceive it as sets of regular correspondences evolved in time'.[32] Here 'evolved' means the unfolding of already present traits, a concept known to naturalists and natural philosophers as preformism.[33] Darwin would later in the *Origin of Species* initially avoid the use of the term evolution because of this preformist association.[34]

What separated Rask from Rydberg and the Grimms from each other was their application of philology to the question of European origins and the degree to which they valorized these origins. This search for origins was at the centre of the Grimms' philological project. In contrast, Rask assumes that the Nordic languages originated in Northern Europe, but that such an origin did not imply an ancient purity. Rask did not concern himself with the finer points of origin, and appears ambivalent regarding the valorization of these origins. There are hints of the folk spirit we find later expressed more directly by Grimm: 'All of the tribes of Gothic offspring, formed in ancient times one great people, which spoke one tongue, that namely which I have now striven to describe'.[35] Grimm noted, though, that the source of their differences lay in the materials they had available to them. Writers such as Rask had an advantage in their work because significant texts had survived in the northern languages.

> We have never had an Edda come down to us, nor did anyone of our early writers attempt to collect the remains of the heathen faith ... Our memorials are scantier, but older; theirs are younger and purer; two things it was important here to hold fast: first, that the Norse mythology is genuine, and so must the German be; then, that the German is old, and so must the Norse be.[36]

What Rask had by way of texts, the Grimms had in the form of spoken language and folk tales.

The Indo-Europeanists forged an apparatus of philological, sociological, psychological, literary, artistic and musical knowledge that in only one hundred years would accumulate to the degree that solution to the Indo-European problem could be found in rationality, bureaucracy, war, internment and industrial murder.[37] The Indo-European problem was never simply about the origins of the Indo-European homeland: but it was also about its possible restoration. Philology and natural history were in fundamental agreement. Together the two fields would 'supply the deficiency, unite the uninterrupted thread of tradition, and by reading the past in the present, reestablish the genealogy of Nations'.[38] Within the study of language the only real challenge to linking philology to folkways came with Saussure, whose rejection of history in favour of structure was a direct refusal of the historical anthropological turn of the comparative linguists. 'First of all, race. It would be a mistake to believe that one can argue from a common language to consanguinity or to equate linguistic families with anthropological families'.[39]

Africa: Filling in the Margins

Africa was the one great deficiency in geography. Not all of Africa, to be sure. European Africa – Northern, Mediterranean Africa, 'the narrow coastal tracts' and the Nile River valley – as Hegel called it, was well known and integrated into the European realm. But the rest of Africa was no more known in Hegel's day than it was in Pliny's, and perhaps less so. The largely unknown geography of Africa plays as important a role in Hegel's characterization of Africans and Africa as it does in Linné and Blumenbach's descriptions of human variations. Except for 'European Africa' the continent 'is almost entirely unknown to us', and 'enveloped in the dark mantle of night'. Periodically terrible hordes emerged due to some 'internal movement' and like a force of nature prevented any social development by ravishing the semi-barbaric coastal plains. Because of them, these regions of Africa have 'no historical part of the World; it has no movement or development to exhibit ... What we properly understand by Africa, is the Unhistorical, Undeveloped Spirit, still involved in the conditions of mere nature, and which had to be presented here only as on the threshold of the World's History'. We have in Hegel's description the nightmarish spectre of suddenly unleashed hordes emerging from the interior of the continent whose 'rage is spent [in] the most reckless barbarism'. Their manner of life and their method of warfare contrasted sharply with the familiar Negroes, those 'peaceful inhabitants of the declivities' or slopes who had adopted some level of civilized life. The domestication brought about by prolonged contact with the civilized peoples renders these few 'mild and well disposed towards Europeans'. Hegel contrasts this Africa to European Africa, a region he says '*must* be attached to Europe', a thin ribbon of the civilized world bounded by a 'girdle of marsh land with the most luxuriant vegetation, the especial home of ravenous beasts, snakes of all kinds – a border tract whose atmosphere is poisonous to Europeans'.[40]

In the *Philosophy of History*, the unknown geography of Africa and its equally mysterious peoples stand in contrast to others little known or understood by the Europeans.

> During the three or three and a half centuries that the Europeans have known this border-land and taken it into their possession, they have only here and there (and that but for a short time) passed these mountains, and have nowhere settled down beyond them. The land surrounded by these mountains is an unknown Upland.[41]

It is worth remembering that, for Pliny, filling in the margins of the world established the fullness of the empire. Hegel went in a different direction and found in the vast emptiness of Africa the proof of a universal history stretching from the earliest times to Enlightenment.

> Africa proper, is far as History goes back, has remained – for all purposes of con-
> nection with the rest of the world – shut up; it is the Gold-land compressed within
> itself – the land of childhood, which lying beyond the day of self-conscious history, is
> enveloped in the dark mantle of Night. Its isolated character originates, not merely in
> its tropical nature, but essentially in its geographical condition.[42]

Hegel elaborates further in his comments on the African. 'The peculiarly African character is difficult to comprehend' because to them the world is fundamentally different from that experienced through the self-conscious understanding of the European. '[I]t does not occur to the Negro mind to expect from others what we are enabled to claim'.[43] The mind of the African lacks even 'the category of Universality' which penetrates all of '*our* ideas'. The African has no sense of individuality, and no sense of will or 'volition'. It is this lack that marks the Negro as a child born of a 'land of childhood' that defines the level of 'Negro life'. The 'Negro exhibits the natural man in his completely wild and untamed state ... want of self-control distinguishes the character of the Negro. This condition is capable of no development or culture, as we see them at this day, such have they always been.' Consciousness, geography character, and religion, all lead to the inevitable conclusion that:

> There is nothing harmonious with humanity in this type ... Fanaticism, which, not-
> withstanding the yielding disposition of the Negro in other respects, can be excited,
> surpasses, when roused, all belief ... Every idea thrown into the mind of the Negro is
> caught up and realized with the whole energy of his will; but this realization involves
> a wholesale destruction.[44]

The Negro now stood at the limit of Enlightenment, just as the Plinian varieties once stood at the edges of the map of the world. This is a very different Hegel than the one described by Buck-Morss as inspired by the revolution in Haiti.[45] The reasons for Hegel's transformation are not exactly clear, but the fact of it cannot be in doubt. One might even expect to find him contrasting Haiti with European Africa, for Haiti was geographically removed from the hordes that prevented the development of the coastal regions. This is not the case and so there is still much work to be done around this question.

Hegel goes on to refer to the 'peculiarly African character' that has no conception of a monotheistic God, just a primitive notion that nature can be influenced through sorcery and ancestor worship. Hegel's Negro recognizes these as evidence of the power of man over nature, but not in a rational sense. It is the dominance of sensual life that prevents the use of reason and the formation of the rule of law. Unable to comprehend the universal, they cannot even understand death as the end of their own individual existence. 'Death itself is looked upon by the Negroes as no universal natural law'. Because Negroes have no recognition of a higher being, 'chance volition' or the use of sorcery has been their power over nature, and this power is made real by the crafting of a 'Fetich'. Conscious of the power of the fetish, the African is never conscious of a higher being represented by it. 'But from the fact that [in Negro religion] man is regarded as the Highest, it follows that he has no respect for himself; for only with the consciousness of a Higher Being does he reach a point of view which inspires him with real relevance'. Lacking any sense of a higher being or a greater law, the Negro is given to seeing the world as arbitrary, and the only absolute truth for them is the arbitrary nature of events. They take their own lives as equally arbitrary and they place no value on their own existence. The development of any real morality is rendered impossible.

> The Negroes indulge, therefore, that perfect *contempt* for humanity, which in its bearing on Justice and Morality is the fundamental characteristic of the race ... Among the Negroes moral sentiments are quite weak, or more strictly speaking, non-existent. Parents sell their children and conversely, children sell their parents as either has the opportunity.[46]

This 'perfect contempt for humanity' is most clearly seen in two aspects of Negro life: tyranny and cannibalism. 'The under valuing of humanity among them reaches an incredible degree of intensity. Tyranny is regarded as no wrong, and cannibalism is looked upon as quite customary and proper. Among us instinct deters from it, if we can speak of instinct at all as appertaining to man.' Just as the instincts and the religious practices of the Negro lack any sense of universality, so too are their political institutions based upon 'mere sensuous volition with energy of will ... There is absolutely no bond, no restraint upon that arbitrary volition. Nothing but external force can hold the State together for a moment. A ruler stands at the head, for sensuous barbarism can only be restrained by despotic power'. The organization of this despotic power as described by Hegel is notably complex for such primitives as he describes. The king is but one of many chiefs and, though he rules, he must take counsel of the others and seek their consent.

> In this relation he can exercise more or less authority, and by fraud or force can on occasion put this or that chieftain out of the way ... If the Negroes are discontented with their King they depose and kill him ... Accompanying the King we constantly find in Negro States, the executioner, whose office is regarded as of the highest consideration, and by whose hands the King, though he makes use of him for putting suspected persons to death, may himself suffer death, if the grandees desire it.[47]

The closeness of the king and the executioner was always noted in European kingdoms as well, but Hegel points to a difference. The politics of the Africans rested on their contempt for humanity and this contempt is characteristic of a slave. 'Another characteristic fact in reference to the Negroes is Slavery'[48] and their tyrannies, cannibalism, 'yielding dispositions' and fanaticism make them perfect slaves. There must be a master, and it makes no real difference whether that master is a chief, a colonial governor or an everyday slave-holder. So we find Africans in the condition of slavery because their contempt for themselves is realized in their contempt for humanity.

It is in slavery itself that Hegel finds the one connection between Europe and Africa.

> The only essential connection that has existed and continued between the Negroes and the Europeans is that of slavery. In this the Negroes find nothing unbecoming them, and the English who have done the most for abolishing the slave-trade and slavery, are treated by the Negroes themselves as enemies.[49]

There are differences, though, in the moral order of slavery in Africa and slavery in America. In Africa, slavery exists as an absolute, whereas in America, the slave-holder exists within larger and stronger moral and ethical circumstances. Slavery in the New World is a civilizing institution and as such, 'we may conclude *slavery* to have been the occasion of the increase of human feeling among the Negroes'.[50] Slavery amongst the Negroes is a manifestation of the same process that Europeans had already passed through. Africans point backwards down the path of the European's own development and represent a primitive moment in the history of the state. It is a moment Hegel's Africans would be condemned to endure indefinitely if not for the intervention and civilizing work of enlightened merchants.

> Negroes are enslaved by Europeans and sold to America. Bad as this may be, their lot in their own land is even worse, since there a slavery quite as absolute exists; for it is the essential principle of slavery, that man has not yet attained a consciousness of his freedom, and consequently sinks down to a mere Thing – an object of no value.

The Negro might even appear at times brave, but it is not true bravery, Hegel says, as it is marked not by the fear of death so much as it is by contempt for life. Whatever noble qualities they display are unintended consequences of their slavery: '[to] this want of regard for life must be ascribed the great courage, supported by enormous body strength, exhibited by the Negroes, who allow themselves to be shot down by the thousands in war with the Europeans. Life has a value only when it has something valuable as its object'.[51] And so this land of childhood cannot be helped, but those transported away under the aegis of the slaver at least had the opportunity to develop the humane ways of civilization:

> existing in a state, slavery is itself a phase of advance from the merely isolated sensual existence – a phase of education – a mode of becoming participant in a higher morality and the culture connected with it. Slavery is in and for itself injustice, for the essence of humanity is Freedom; but for this man must be matured. The gradual abolition of slavery is therefore wiser and more equitable than its sudden removal.[52]

If slavery is a fundamental moment of civilization then the memory of it – and this is what makes slavery a particularly contemporary problem – still weighs upon us and burns its mark regardless of the role our particular ancestors played in it. The common wounds are repressed, but the memory always appears in the particular, in the events and chance encounters of everyday life.[53] As for Hegel, one need look no further for the repression of the wounds of slavery than his concluding remark: 'At this point we leave Africa, not to mention it again.'[54]

Monogenesis and the Fixity of Species

Linné's classification epitomized the arguments for the order of nature while it acknowledged diversity and variation in the elaborate tables of the *Systema Naturae*. In order to achieve a level of scientificity, a system of classification was absolutely necessary. The energies of the great naturalists like Linné, Buffon and Blumenbach were devoted to two integrated lines of inquiry: the origins of species and the meaning of variation in the natural world. The species question had to be resolved before any general scheme of classification and order could be finally established. The conclusion of the species question would decide the admissibility of the study of life – biology and ecology, for example – into the domain of science. The possibility of achieving a truly scientific classification rested in the end upon the ability to establish names that could be understood across borders and over time. The species question concerned the meaning of variety in nature, but such a general theory has always been beyond the reach of naturalists. In fact, such general theories appeared at times more theological than scientific. If the variation found in humans could be understood, then we would have the key to understanding variation throughout nature. By assuming each variety of human on a specific continent and associating it with the flora and fauna of its locale, classifications such as Blumenbach's show us the importance the solution of the species question held for naturalists of the time and the centrality of the new category of race to the ability of naturalists and political philosophers to provide answers crucial for governing these newly racialized populations and lands. That these disciplines were not equal to the task did not diminish the enthusiasm for racial theories in the later sciences of society and life. Thus it was that the problem of the relationship of three new objects of investigation – man, society and the natural world – arose within the more limited species question.

Linné classified the types of humans within a general framework system of nature. In the second edition of his *Systema Naturae*, Linné included humans – the animals with the imperative to know themselves – in the class *Quadrupedia* and the order Anthropomorphia. Within this order Linné noted four varieties of Homo distinguishable by colour and geographic distribution (Table 1.1).

Of note is the simplicity of the scheme and its terse descriptions. Here we can say that language has indeed fallen away only to emerge again in Linné's later classifications.[55] For what else can we say of the enlarging schemes given over the course of the various editions of the *Systema Naturae* except that there is an elaboration of the scheme of human variation? The tenth edition of 1758–9 – the last by Linné – gives a more exacting scheme than the 1740 second edition (Table 1.2). Most obvious is the accumulation of language, the degree of detail added during the intervening eighteen years and the switch in nomenclature from *Quadrupedia* to *Mammalia*. The addition of characteristics that are not physical, including typical garments and stereotypical behaviour, demonstrate an implied correlation between culture and physique. Close vestment and governance by law, for example, are characteristics of advanced civilization and of the most advanced peoples.[56]

Table 1.1: Linné's classification of human varieties as given in *Systema Naturae*, 2nd edn (1748), p. 3.

Classis I
QUADRUPEDIA
Ordo 1
Anthropomorpha. *teeth* four fore-teeth, or none
1. Homo. Know thyself

| Homo varieites: | Eurpaeus albus | Asiaticus fuscus |
| | Americanus rubescens | Africanus niger |

Linné's table is internally more complex and descriptive in the tenth edition, including a more detailed physical description along with psychological ('Gentle, acute, inventive'), cultural ('Covered with loose garments') and social ('Ruled by opinions') descriptions as well. While no hierarchy need exist, there is a steady movement that can be read either as a progression from the primitive ('Anoints himself with grease') to the culturally sophisticated ('Covered with close vestments'), or in the downward movement of the civilized '(Governed by laws') into the degenerate ('Governed by caprice'). It is precisely the flexibility of the scheme that allows it to be read in either direction – as progress or degeneration – and made it such a powerful system of classification and value.

This scheme worked in the opposite direction from the break that is said to exist between natural history and biology.[57] Instead of the circumscription of the range of characteristics and the removal of language (in the form of stories, folk tales, etc.) from consideration, what we find in the cases of Linné and Blumenbach is the constant layering on of additional language, or at least the accumulation of new discourses which explain and mark the differences between types or varieties of humans. A scheme of classification laid out in the manner of the early editions did nothing to explain how one might understand the differences between humans and between apes and humans. Writing his dissertation under Linné's direction, Christian Emmanuel Hoppe addressed the central problem that confronted natural history: the similarity between apes and humans.

Table 1.2: Linné's classification of human varieties, adopted from *Systema Naturae*, 10th edn (1758–9), pp. 20–2.

Mammalia

I. Primates
foreteeth, upper 4, parallel
Petoral mammae, 2
I. *Homo Sapiens* know thyself
1. H. diurnus. Man of the day. roaming cultivation, places

Ferus	1. varying by culture and place on all fours, mute, hairy
Americanus	a. H. rufus reddish, choleric, erect
	Hair black, straight, thick; *Nostrils* wide; *Face* harsh, *Beard* scanty
	Obstinate, merry, free
	Paints himself with fine red lines
	Regulated by customs
Europaeus	b. H. albus white, sanguine, muscular
	Hair flowing, long, *Eyes* blue
	Gentle, acute, inventive
	Covered with close vestments
	Governed by laws
Asiactus	c. H. luridus sallow, melancholy, stiff
	Hair black. *Eyes* dark
	Severe, haughty, avaricious
	Covered with loose garments
	Ruled by opinions
Afer	d. H. niger black, phlegmatic, relaxed
	Hair black, frizzled. *Skin* silky. Nose flat. *Lips* tumid
	Women without shame. *Mammae* lactate profusely
	Crafty, indolent, negligent
	Anoints himself with grease
	Governed by caprice
H. monstrosus	e. Lives alone (a), some arts (b,c)
	a. *Alpini* small, child-like, agile, fearful; *Patagonici* large, sluggish
	b. *Monorchides* little beasts: Huttentotti
	Juncea girls with large belly: European
	c. *Macrocephali* coneheads. China
	Plagipcephali flatheads. Canada

2. H. Nocturnus. Man of the night. Ourang Outang ...

> I can discover scarcely any mark by which man can be distinguished from the apes ...
> I am still uncertain, by what characteristic mark the troglodytes can be distinguished
> from man, according to the principles of Natural History. For there are so many
> things so alike in these kinds of apes and man, such as the structure of the almost bare
> body, the face, the ears, mouth, teeth, hands, breasts; and also in the food, imitation,
> and gesticulations in those species which walk upright and are properly called anthro-
> pomorpha, that it is very difficult to find marks sufficient to divided the genus.[58]

As a teacher, it was Linné's practice that dissertations were literally dictated to
the student, who then reworked their transcripts of these personal lectures into
manuscript form. In the dissertations of Linné's students, we can be reasonably
sure that their views did not deviate much from those of their teacher.[59] Hoppe's
writing on human variety closely follows similar discussions in Linné's works.
Another student writing his dissertation with Linné noted that:

> it is for us to inquire what man has beyond other animals. Anatomy teaches us that
> man possesses heart, brain, entrails, nerves, bones, muscles; that he moves himself,
> touches, tastes, smells, hears, and sees, exactly like brute animals. It is, however, true
> that you will find two peculiarities in man, of which the other animals are destitute,
> – things that we do, but being destitute of any distinct perception, they cannot apply
> reflection to any objects whatsoever, or that attention of looking at anything, which
> a man does when absorbed in admiration.[60]

For Linné, everything in the world was connected by a great chain of being,
gradually increasing in complexity from simple species at one end to our own at
the other end of 'this natural chain'. Because of this chain of being, we are both
a part of nature and the one species that can understand nature and ourselves
through the use of Reason. 'It is reason in which man is pre-eminent. In no other
faculty does he excel other animals ... Man is distinguished from all other animals principally by reason.' Humans are a 'miracle of nature's audacity. The chief
of animals, for whose sake nature has produced everything.' Their qualities are
many, but so too are their less attractive qualities equally abundant: 'weeping,
laughing, singing, docile, judging, wondering, very wise, but delicate, naked,
defenseless by nature, exposed to all the contempt of fortune, dependent upon
the assistance of others, of anxious mind, and desirous of protection, of wavering
spirit, obstinate in hope, querulous in life, very slow in gaining wisdom'.[61]

Still, the connection between humans and apes was clear to Linné. In 1747,
he wrote 'I demand of you and of the whole world, that you show me a generic
character – one that is according to generally accepted principles of classification
by which to distinguish between Man and Ape. I myself most assuredly know
of none.'[62] Linné's view that *Homo troglodytes* was a link in the chain between
humans and 'the nearest relations of the human races' would have many consequences. Man was placed in relation to other animals, but exactly what this
relation was and how it was to be used to make sense of human variety could

not be resolved by taxonomy and a unified classification remained an aspiration. Here, the Hottentots took their place in the interstice between man and ape.

> This preoccupation with the question of man's relation to the anthropods gave an especial 'philosophical' interest to the rather numerous descriptions of the Hottentots by the late seventeenth- and early eighteenth-century voyagers. They were probably the 'lowest' race thus far known; and more than one writer of the period saw in them a connecting link between the anthropods and homo sapiens.[63]

A generic morphological character upon which to fix variety was not to be found, but differences of rationality and culture lay on the surface. The scheme of classification expanded its categories until the final tenth edition and always relied as much on cultural as it did on morphological differences. In defending his view that humans should be classified along with the quadrupeds, Linné argued that reason and speech are the only attributes that separate us from the animals, but, in his scheme, these same attributes also separate us from each other.

For all the rigidity of his binomial nomenclature and his belief in the fixity of species in terms of both number and kind, Linné was flexible where it came to the question of human variety. The ape shades into man, and man into the ape. A degree of flexibility that, in other words, did not require the renunciation of the chain of being. The innovation of placing man in the natural order has been justly noted ever since the first edition of the *Systema Naturae*, but the monstrous and the great apes were both included in the same order as the human.[64] Less noticed was the expansive notion of human difference in the successive revisions of Linné's scheme. Having established these differences, it now remained to identify the origin of the differences between humans.

Monogenesis, Infinite Variety and Degeneracy

Within natural history, the monogenic versus polygenic debates were always embedded in the debates on the species question. The species question was a broader concern, but both confronted similar issues and addressed the same objects of study. One could be, like Buffon, a monogenist and also believe in the fixity of species. 'Every species having been originally created, the first individual served as a model to their descendants ... the existence of species is constant'.[65] This was perhaps Buffon's most important contribution to the species question. The fixity of species does not limit the variety of nature in the sense that the number of species might reach to infinity, but the individual species created maintained forever a fixed design. These fixed species expressed the organization of nature through anatomy, behaviour, functions and the relationships between species. If all these could be put in their geographic locale, then the taxonomic table would represent the dense and hierarchical system of nature.

> Buffon ... sought to encompass all the overt complexity of organisms into a non-hier-
> archical system that recognized differing relationships for various properties (bats,
> for Buffon, stood closer to mammals in anatomy, and closer to birds in function). But
> this alternative model of a network with multiple linkages, rather than a strict hierar-
> chy of inclusion, fails (in the admittedly retrospective light of evolution) to separate
> the superficial similarity of independent adaptation (wings of bats and birds) from
> the deep genealogical linkages of physical continuity through the ages (hair and live
> birth of bats and bears). Buffon's noble vision of equal treatment for all aspects of a
> species' life – placing ecology, function, and behavior at par with traditional anatomy
> – foundered on a false theory about the nature of relationships.[66]

This vision of multiplicity in nature can be found in the specific causes of human
variety, for anatomical variety emerged alongside variations in climate, food and
manners although the effects of climate gave it great influence. Climate did not
change. The climatic zones remained an assumption of naturalists. Much attention has been given to the debate over the age of the earth and its changes in terms
of geology, but the naturalists of the time lacked an idea of climate shift even
under the influence of Cuvier's revolutions of the earth. Climate was believed
to be the first cause, as Buffon observed: 'The climate may be regarded as the
chief cause of the different colors of men. But food, though it has less influence
upon color, greatly effects the form of our bodies. Coarse, unwholesome, and ill
prepared food, makes the human species degenerate'.[67] The potential for degeneracy in society and the possibility that human variety could itself be the result of
degeneration would become commonplaces in the century to come. But Buffon
opposed producing a hierarchical classification based upon this notion of degeneracy. Despite the obvious variety of types and anatomical differences which he
took to be pronounced forms of degeneracy, Buffon held that the monogenic
origins of humans could not be denied. We remain a single species as proven by
the interfertility of the races.

> The Asiatic, the European, and the Negro produce equally with the American. Noth-
> ing can be stronger proof that they belong to the same family, than the facility with
> which they unite to the common stock. The blood is different, but the germ is the
> same. The skin, the hair, the features, and the stature, have varied, without any change
> in internal structure.[68]

If interfertility expressed the common 'internal structure' of all humans, why
then are humans divided into seemingly distinct varieties and why is there a
resemblance in body and behaviour to other species, especially the apes? Savage and Wyman grappled mightily with these issues in their 1847 report on an
'unknown species' brought back to the United States for dissection and investigation. They named it *Troglodytes gorilla*, now known simply as a gorilla,

a term used by Hanno in describing the 'wild men' found on the coast of Africa ... The organization of the anthropoid Quadrumana justifies the naturalist in placing them at the head of the brute creation, and placing them in a position in which they, of all the animal series, shall be the nearest to man. Any anatomist, however, who will take the trouble to compare the skeletons of the Negro and the Orang, cannot fail to be struck at the sight of the wide gap which separates them. The difference between the cranium, the pelvis, and the conformation of the upper extremities in the Negro and Caucasian, sinks into comparative insignificance when compared with the vast difference which exists between the conformation of the same parts in the Negro and the Orang. Yet it cannot be denied, however wide the separation, that the Negro and Orang do afford the points where man and brute, when the totality of their organization is considered, most nearly approach each other.[69]

Confronted with the body of a different species of primate, Savage and Wyman acknowledge that their examination of the gorilla's corpse showed the 'insignificance' of the differences between humans. But we may rest assured that the totality of their organization makes Negroes a part of the chain where 'man and brute' are connected, but where Caucasian remains as distant from the Orang as from the Negro. Although the explanation for this contradiction had to be offered without the benefit of our modern concepts of natural selection, fitness, genetic drift and reproductive isolation, explanations put forward resembled these more familiar concepts. Adaptation to a specific geographical and climatic locale proved to be a force behind variation only as far as the dogma of fixity allowed.

> Upon the whole, every circumstance concurs in proving, that mankind are not composed of species essentially different from each other; that on the contrary, there was originally but one species, who, after multiplying and spreading over the whole surface of the earth, have undergone various changes by the influence of climate, food, mode of living, epidemic diseases, and the mixture of dissimilar individuals; that, at first, these changes were not so conspicuous, and produced only individual varieties; that these varieties became afterwards specific, because they were rendered more general, more strongly marked, and more permanent, by the continual action of the same causes; that they are transmitted from generation to generation, as deformities or diseases pass from parents to children; and that, lastly, as they were originally produced by a train of external and accidental causes, and have only been perpetuated by time and the constant operation of these causes, it is probable that they will gradually disappear, or, at least, that they will differ from what they are at present, if the causes which produced them should cease, or their operation be varied by other circumstances and combinations.[70]

Given the influence of climate, diet, geography, etc., change from the original forms – which Buffon termed degeneration – made systematic classification impossible. His belief in change as degeneration caused by isolation and environment led to a scientific confrontation with Thomas Jefferson over the source of variety amongst North American flora and fauna.[71]

Thomas Jefferson's only full length book, *Notes on the State of Virginia*, critiques Buffon's argument for the uniqueness of New World species – i.e., that they are degenerate versions of Old World species. A work of natural history, the *Notes* develops the theme that the scientific meaning of variety has political ramifications in the New World that are explicitly linked to problems of civilization and government. Jefferson's discussion of Virginia and North America show his debt to and his critical reading of Linné and Buffon. In particular, the chapter 'Productions Mineral, Vegetable, and Animal' is an explicit critique of Buffon and owes much to Jefferson's reading of Linné's nomenclature. An 1814 letter by Jefferson regarding the question of classification provides further evidence of his interest in the subject. A Dr John Manners had inquired regarding Jefferson's views on the many competing methods of classification. Jefferson begs off a definitive reply, noting that he can only speak generally and regarding only those naturalists who had influenced his own scientific endeavours. Although 'a life of continual occupation in civil concerns has so much withdrawn me from studies of that kind', Jefferson proceeded to demonstrate his familiarity with the issue by giving a fairly lengthy and detailed response. While his correspondent had noted that disease had become a unit for measuring nature, Jefferson took issue with the notion of degeneracy as a starting point for his discussion of the exceptionalism of American fauna and flora:

> you say that disease has been found to be an unit [of measure]. Nature has, in truth, produced units only through all her works. Classes, orders, genera, species, are not of her work. Her creation is of individuals. No two animals are exactly alike; no two plants, nor even two leaves or blades of grass; no two crystallizations. And if we may venture from what is within the cognizance of such organs as ours, to conclude on that beyond their powers, we must believe that no two particles of matter are of exact resemblance. This infinitude of units or individuals being far beyond the capacity of our memory, we are obliged, in aid of that, to distribute them into masses, throwing into each of these all the individuals which have a certain degree of resemblance; to subdivide these again into smaller groups, according to certain points of dissimilitude observable in them, and so on until we have formed what we call a system of classes, orders, genera and species. In doing this, we fix arbitrarily on such characteristic resemblance's and differences as seem to us most prominent and invariable in the several subjects, and most likely to take a strong hold in our memories. Thus Ray formed one classification on such lines of division as struck him most favorably; Klein adopted another; Brisson a third, and other naturalists other designations, till Linnaeus appeared.

Linné had presented a system of three kingdoms 'which had obtained the approbation of the learned of all nations ... [and] his system was accordingly adopted by all, and united all in a general language'. The *Systema Naturae* was superior to earlier classifications because it alone aided 'the memory to retain a knowledge of the productions of nature'. It made it possible at long last to give 'the same names for the same objects, so that [naturalists] might communicate understandably on them'. Finally, Linné's system allowed one to take a new specimen and 'trace it by its character up to the conventional name by which it was agreed to be called'. The confusion amongst naturalists regarding classification and nomenclature had been solved, Jefferson wrote, but even now Linné's 'disciples' – along with those of Blumenbach and Cuvier – had pressed their respective systems so far that naturalists had returned to a pre-Linnean chaos, with naturalists producing their own idiosyncratic and mutually unintelligible systems and nomenclatures. In the end, for all of his praise of Linné and respect for Blumenbach and Cuvier, Jefferson stresses that systems of classification are not natural but imposed by humans in order to understand nature rationally. Even Linné's system was

> liable to the imperfection of bringing into the same group individuals which, though resembling in the characteristics adopted by the author for his classification, yet have strong marks of dissimilitude in other respects. But to this objection every mode of classification must be liable, because the plan of creation is inscrutable to our limited faculties. Nature has not arranged her productions on a single and direct line. They branch at every step, and in every direction, and he who attempts to reduce them into departments, is left to do it by the lines of his own fancy.[72]

Jefferson offers much praise for Buffon, though. The tables of classification that Jefferson included in the *Notes* owe their organization and authority to Buffon: 'I take him for my groundwork, because I think him the best informed of any Naturalist who has ever written'.[73] However, Buffon's view of monogenesis and gradual differentiation within fixed limits, while generally supported by Blumenbach, was to be overcome by linking Blumenbach's system of racial classification with Cuvier's catastrophe theory of 'successive creations'.[74] Jefferson's views ultimately aligned more with Buffon than Linné on the issue of the fixity of species. Indeed, to prove Buffon wrong regarding the degenerate origins of New World species, he drew on Linné to establish the relative fixity of species, but in so doing argued that the concept of the variety of nature that Buffon used against Linné could just as easily be used as an argument against Buffon:

> The opinion advanced by the Count de Buffon, is 1. that the animals common both to the old and the new world, are smaller in the latter. 2. That those peculiar to the new, are on a smaller scale. 3. That those which have been domesticated in both, have degenerated in America. 4. That on the whole it exhibits fewer species.

Jefferson meets each of these assertions, often by first appealing to his own direct experience, and then to a general theoretical position, for 'when we appeal to experience, we are not satisfied with a single fact'. The first two assertions regarding size and weight were easily disproved using comparative tables of New and Old World species. The third assertion is more important as it is the one that is 'applied to brute animals' and by 'extension to the man of America, whether aboriginal or transplanted'. Jefferson argues against what he found to be a fundamental error in Buffon's view of Native Americans. Quoting at length from Buffon's text, Jefferson carefully refutes each point given in support of degeneracy in a detailed defence of 'Aboriginals and transplants' but scarcely anything is mentioned by Jefferson regarding the possible degenerative effects of slavery, or of the New World on slaves. For Jefferson, the fate of the Negro and the fate of the Native American were not connected in any sense and he is ambivalent regarding the possible cultivation of 'the Black Man'.

> I believe the Indian, then, to be, in body and mind, equal to the white man. I have supposed the black man, in his present state, might not be so; but it would be hazardous to affirm, that, equally cultivated for a few generations, he would not become so. As to the inferiority of the other animals of America, without more facts, I can add nothing to what I have said in my *Notes*.[75]

At times Jefferson described his direct observations of Native Americans as occurring over a longer period and being of greater mutual benefit than his experiences with Negroes and suggested that his observations on the former are more reliable than on the latter. He had devoted a detailed chapter of the *Notes on the State of Virginia* to its aborigines. He is impressed by what he sees as the lack of 'any shadow of government. Their only controls are their manners, and that moral sense of right and wrong, which like the sense of tasting and feeling, in every man makes a part of his nature.' Offences are punished by 'contempt' or 'exclusion' while more serious offences are punished 'by the individuals whom it concerns'. When compared to the lawless Americans the government of 'civilized Europeans ... submits man to the greatest evil ... the sheep are happier of themselves, than under the care of the wolves'.[76] We find this same early concern for the human rights of Native Americans in the appendix to the *Notes*, in which he published an account 'Relative to the Murder of Logan's Family'. In events that led to Dunmore's War, a Virginia militiaman ordered a massacre that included the family of Logan, a Shawanese chief. In his chapter 'Productions Mineral, Vegetable, and Animal' Jefferson refers to Logan in order to prove that those Indians whom he had himself encountered were capable of being equals. 'I may challenge the whole orations of Demosthenes and Cicero, and of any more eminent orator, if Europe has furnished more eminent, to produce a single passage, superior to the speech of Logan, a Mingo chief, to Lord Dunmore'. On the 'great question' of the origins of America's 'aboriginal inhabitants' Jefferson believes that they came either from Asia across the Bering Sea or, despite 'the imperfect navigation of ancient times', from Europe. The proof of these migrations would be found in the languages of the Americas, but already in his time too many tribes had been extinguished to make such a reconstruction possible.

It is to be lamented, very much lamented, that we have suffered so many of the Indian tribes already to be extinguished, without our having previously collected and deposited in the records of literature, the general rudiments at least of the languages they spoke. Were vocabularies formed of all the languages spoken in North and South America, preserving their appellations of the most common objects in nature, with the inflections of their nouns and verbs, their principles of regime and concord, and these deposited in all the public libraries, it would furnish opportunities to those skilled in the languages of the old world to compare these, now, or at a future time, and hence construct the best evidence of the derivation of this part of the human race.[77]

Jefferson referred Chastellux to his refutation of Buffon in the chapter on 'Productions Mineral, Vegetable, and Animal'. Though one might expect a naturalist of his day to include here a full treatment of degeneracy and slavery, instead one finds these discussed in the chapter on the 'Laws of the Commonwealth of Virginia'. Elsewhere in the *Notes* Jefferson only mentions Negroes in passing, saving the majority of his comments for the later section on law, as might signal an understanding of slavery as a legal condition as opposed to a natural condition, but an interest in blacks only in terms of their legal condition. While he could clearly state that the enslavement of Indians was 'inhuman', we do find such a definitive statement regarding Negroes but nor do we find a real defence of Negro slavery.

In an echo of the importance of the monstrous in natural history, Jefferson does pause briefly to 'add an account of an anomaly of nature', sometimes found in the race of Negroes brought from Africa, who, though black themselves, have in rare instances, white children, called albinos. 'I have known four myself and have faithful accounts of three others'. This digression is added to entertain the reader and follows the convention of natural history writing on 'sports' and monstrosities, or in Jefferson's preferred term 'anomalies'. But does it have one other significance, coming as it does in a chapter in which he otherwise ignores human variety, and he ends by writing that this 'anomaly of nature' must be added to the 'catalogue of our indigenous animals'. The term albino itself covered a wide range of types, including mulattoes and those who lose their pigment through life as well as those who are born a 'pallid cadaverous white' though with 'no mixture of white blood'. The defect was not a familial one, nor did it appear contagious. Their own 'issue was black', but all were 'uncommonly shrew, quick in their apprehensions and in reply'. Jefferson announced that 'whatever the cause of the disease of the skin, or in its coloring matter, which produces this change, it seems more incident to the female than male sex'. Jefferson also relates the case of another type of albino,

a Negro man within my own knowledge born black, and of black parents; on whose chin, when a boy, a white spot appeared. This continued to increase till he became a man, by which time it had extended over his chin, lips, one cheek, the under jaw and neck on that side. It is of the Albino white, without any mixture of red, and has for several years been stationary. He is robust and healthy, and the change of colour was not accompanied with any sensible disease, either general or topical.[78]

Albinos were of great interest to many naturalists and even today are still seen in significant portions of the world as monstrosities to be killed out of fear or as sources for folk medicine or both. Linné included them in his monstrous category as a type of African 'troglodytes' and Voltaire wrote of one in his essays highlighting the practice of putting these humans, usually having been taken as slaves, on public display.[79]

Despite this vestige of the medieval focus on monstrosities, Jefferson exposed the political questions which cut through natural history by dividing his discussion on human variety between his chapters on economic production and on the legal system. Obviously Jefferson's legacy is not so much in his passion for science but in his work as a politician and statesman. Even if his own presence was lacking in most scientific disputes, the discoveries and critiques which he discussed were common knowledge among the educated antebellum classes. His affinity for Enlightenment rationality was clear from his early writings on Christianity, such as his *Life of Jesus*, where he deletes the religious aspects of the New Testament, leaving only a real-world ethics. He noted that a 'system thus collected from the writings of Tully, of Seneca, or Epictetus, and others, would be more full, more entire, more coherent and more deduced from unquestionable principles of knowledge'.[80] In fact, Jefferson's decision to place the discussion of slavery in a separate chapter on law, rather than in his discussion of the natural history of the Virginia colony, may signal somewhat critical attitude towards polygenic views. Most telling, Jefferson did not attempt to separate questions of origin, progress and degeneracy from questions concerning the legality and justice of slavery. For the American School, these questions are all one and the same. As we shall see in the next chapter, their answer that one's origins determines one's freedom would be radically different.

The American School extended and refined Jefferson's critique of Buffon's view of degeneracy, but they adopted a concept of fixity and of a fixed system of classification Buffon and Jefferson had at the very least avoided.

That all forms of life in this country were wanting in vigor, and generally inferior to those of the eastern continent, was maintained by Buffon, De Pauw, and the Abbe Raynal, and partially adopted by Robertson. It was indignantly repelled by Jefferson, and is termed by Morton an 'idle theory', and an hypothesis of 'closet naturalists', which there is ample evidence to disprove. An explanation of the assumed fact was sought by its supporters in the supposition that this continent emerged from the water [of the biblical flood] at a later period than the other, and had not recovered from the effect of cold and moisture, which exerted an enervating influence upon the inhabitants, whose resemblance and uniformity showed them to be more recent than the people of the other hemisphere, and that time had not been afforded them to become as robust as the latter.[81]

Classification, Catastrophe and Extinction

Professor of medicine at the University of Gottingen and archivist of skeletons and crania from across the globe, Blumenbach was reputed to possess the largest collection of crania in the world until surpassed by Samuel Morton in Philadelphia. It was, however, on the basis of just one skull from this collection that Blumenbach coined the term *Caucasian* to describe a human type that he said was the ancestor of all modern humans: the European type. Blumenbach identified the Caucuses Mountains as the original home of the Europeans in general and of the Germans in particular. That this original home lay to the east of Germany and would often appear later in notorious propaganda is nothing that Blumenbach would have intended, still it cannot be ignored that such views were a part of the same tendency found in the philologists, genealogists and geographers searching for the homeland of the Aryan.

As a collector Blumenbach also amassed the largest collection then known of works by authors of recent African descent, and championed the poetry of Phyllis Wheatly (in whom Jefferson found only what he called 'the spirit of religion', but not 'the spirit of poetry').[82] He rejected the concept of a great chain of being and understood humans as so unique that we deserved our own order separate from other mammals, even if this classification came at the expense of breaking the great chain of being. Blumenbach's racial classification closely linked variety to geographical locale. Blumenbach noted that even the most 'primitive' race is so varied within itself that one might

> easily take the inhabitants of the Cape of Good Hope, the Greenlanders, and the Cicassians for so many different species of man, yet when the matter is thoroughly considered, you see that all that do so run into one another, and that one variety of mankind does so sensibly pass into the other, that you cannot mark out the limits between them ... of the different varieties ... it has been asserted that the Negroes are specifically different in their bodily structure from other men, and must be placed considerably in the rear, from the condition of the obtuse mental capacities. Personal observation, combined with the accounts of trustworthy and unprejudiced witnesses, has, however, long since convinced me of the want of foundation in both these assertions.[83]

Only by rupturing the chain of being could Blumenbach assert the unity of human varieties. A continuum from apes to human permitted one to think humans as part of the natural world, but it also allowed one to think of the ape as close to the Negro, and far from the European.[84] Ironically, monogenesis could at this time be scientifically asserted only by reducing the ape to being a primitive in a preformist sequence that moved inevitably towards humanity.

There is a continuity in classifications from Linné, Blumenbach and later commentators on the scientific importance and permanence of racial characteristics. The theory of the fixity of species easily aligned with – and could be transformed into – the fixity of racial characteristics.[85] Blumenbach had acknowledged that the classifications of human variety are '[v]ery arbitrary indeed both in number and definition have been the varieties of mankind accepted by eminent men'.[86] The arbitrary nature of classification could be found in the sheer variety of classifications and in the obvious variety across individuals of a single species. He would amplify his view in the third edition of his *De generis humani varietate nativa* with the implicit assumption that greater variety is found in species of greater age and that the older the species, the more its individuals will vary from each other. Modification could be acknowledged across all humans and still support the versions of the monogenic theory which Blumenbach defended. Far from denoting separate of origins, variety could clearly establish the essential unity of humans: 'no variety of mankind exists, whether of colour, countenance, or stature, etc., so singular as not to be connected with others of the same kind by such imperceptible transition, that it is very clear that they are all related, or only differ from each other in degree'. The systems of classification 'accepted by eminent men' are arbitrary. They are only useful for organizing facts because it is 'serviceable to the memory to have constituted certain classes into which the men of our planet may be divided'.[87] One might ask how, given his ambivalence, Blumenbach's work would come to establish the basic system of classification of human variety that we have today.

Blumenbach particularly objected to using aesthetics as a basis for classification. While he certainly believed his 'Caucasian' specimen to be the most beautiful, the meaningful classification of human varieties remained a scientific and not an aesthetic question. Blumenbach reminded those who looked to aesthetics that 'If a toad could speak and were asked which was the loveliest creature on God's earth, it would say simpering, that modesty forbade it to give a real opinion on that point'.[88] A new system of scientific classification would surely be of great assistance against Camper and others who classified and ranked varieties in accordance with aesthetic principles. In his first classification Blumenbach followed closely the system given by Linné in the second edition (1740) of the *Systema Naturae* and presented only four varieties of human. His classification differed from Linné in his coining of the term Caucasian and in his addition of the new category of Malay. We might pause to note that these innovations are not found in the first edition (1776) of his *De generis humani*. In the earlier edition, Blumenbach classified humans into four races of one species, but the later expanded classification offered nothing less than the desired 'physio-geographical history of man'.[89] The variety of humankind mapped out Blumanbach's understanding of the variety of the earth and how humans could be included in its geographical description:

Caucasian (Europe, white), Mongolian (Asia, yellow), Ethiopian (Africa, black), American (Americas, red) and Malayan (Pacific, brown).

> The first and most important to us (which is also the primitive one) is that of Europe, Asia this side of the Ganges, and all the country situated to the North of the Amur, together with that part of North America, which is nearest both in position and character of the inhabitants. Though the men of these countries seem to differ very much amongst each other in form and colour, still when they are looked at as a whole they seem to agree in many things with ourselves. The second includes that part of Asia beyond the Ganges, and below the river Amur, which looks towards the south, together with the islands, and the greater part of those countries which are called Australia. Men of dark color, snub noses, with winking eyelids drawn outwards at the corners, scant, and stiff hair. Africa makes up the third. There remains finally, the fourth, the rest of America, except so much of the North as was included in the first variety.[90]

The classifications of Linné and Blumenbach correlated skin colour with geographical regions but these geographical places were more real that the racial categories. In fact, they were constructed in much the same manner, with the presence of the racial type defining the geographical domain at least as much as geographical notions set the foundation and limits of racial types. *Africus* is where one finds *niger*, and where one finds *niger* there too is *Africus* and, like Linné's, Blumenbach's table becomes more complex over time.[91] The further away or down you descend on the table, the less variety there is. Asians are described but Africans and Americans are not. The principle of gradation is embedded in the table itself.

In the second edition of *De generis humani* (1781), Blumenbach presented us with the now familiar five races (Table 1.3), and gave an increasingly detailed listing of the characteristics of each. Breaking with Linné, Blumenbach writes that his division of humans into five races instead of four was the result of his intensive research into 'the different nations of Eastern Asia and America'. Separating humans into five varieties was:

> more consonant with nature. The first of these and the largest which is also the primeval one, embraces the whole of Europe, including the Lapps, whom I cannot in any way separate from the rest of the European, when their appearance and language bear such a testimony to their Finnish origins ... All these nations regarded as a whole [Blumenbach includes in Europe the 'western part of Asia ... also northern Africa, and ... the Greenlanders and Esquimaux'] are white in color, and, if compared to the rest, beautiful in form.
>
> The second variety comprises the rest of Asia ... The inhabitants of this country are distinguished by being brown in color, more or less verging on olive, straight face, narrow eyelids, and scanty hair ...
>
> The third variety comprises what remains of Africa ... Black men, muscular, with prominent upper jaws, swelling lips, turned up nose, very black hair.
>
> The fourth comprises the rest of America, whose inhabitants are distinguished by their copper color, their thin habit of body, and scanty hair.

Finally, the new southern world makes up the fifth ... the men throughout being of a very deep brown color, with broad nose, and thick hair[92]

Blumenbach replaced Linné's *Africus niger* with *Ethiopian*, *Europaeus* with *Caucasian*, and *Asiaticus* with *Mongolian; ferus* and *monstrosus* drop away and *Malayan* is added. The great chain of being may have been a notion that Blumenbach abhorred, but it was a notion which coexisted with the view that the different varieties and the different locales blend together and express the range of real and potential variation.[93] Blumenbach preserved this aspect while severing the chain itself with the division of humans into five races.

Table 1.3: **Blumenbach's classification, in** *De generis humani varietate nativa*, 2nd edn (1781), pp. 269–76.

Name	Geographical Location	Colour	Characteristics
Caucasian	Europe, including Lapps, North Africa, America, Eskimo and Greenlanders derived from Lapps, Western Asia	White	Beautiful in form
Mongolian	Rest of Asia	Brownish/ Olive	Straight face, narrow eyelids, scanty hair
Ethiopian	Africa excluding Northern Africa	Black	Muscular, prominent upper jaws, swelling lips, upturned nose, very curly black hair
American	Non-European Americans	Copper	Broad nose, scanty hair, thin habit of body
Malayan	Southern Pacific	Very deep brown	Broad nose, thick hair

Moreover, Blumenbach's system represented more than a simple 'transposition from the Linnean geography to hierarchical ranking'.[94] Both Linné's and Blumenbach's systems were based on geographical distribution and an implicit hierarchy, though Linné was less interested in this than Blumenbach.

> The shift from a geographical to a hierarchical ordering of human diversity marks a fateful transition in the history of Western science – for what, short of railroads and nuclear bombs, had more practical impact, in this case almost entirely negative, upon our collective lives and nationalities. Ironically, J. F. Blumenbach is the focus of this shift – for his five-race scheme became canonical, and he changed the geometry of human order from Linnaean cartography to linear ranking by putative worth.[95]

The concept of race in its most modern form united the vastly different depictions of human diversity by Blumenbach over the many editions of the *De generis humani*. The races are presented in a system 'that placed a single race at the pinnacle of the closest approach to the original creation, and then envisioned two symmetrical lines of departure from this ideal toward greater and greater degeneration.'[96]

Blumenbach presented his cranial series in two very different illustrations; one invites a spatial and the second a hierarchical interpretation of his classification. Although deviation did not mean differences in the degree of humanity, Blumenbach's revision of his system expressly ties racial categories to geographical ones. The races are arranged in a pyramid-like order in one, but as a horizontal series in the other. Blumenbach considered the Caucasian to be the original from which all others had degenerated. Likewise, Linné's system certainly allows one to read it in terms of a hierarchy or hierarchical chain of being. Blumenbach's system is distinguished by the lack of any fanciful categories. Linné had included two varieties of *Homo sapiens* which Blumenbach rejected: *ferus* and *monstrosus*. *Ferus* designated those children raised in the wild, who lacked culture but whose physical attributes were clearly humanoid. The *monstrosus* denoted those monstrous races that we inherited from Pliny, his sources and his medieval commentators. The inclusion of *ferus* and *monstrosus* places Linné in the period when the monstrous still dwelled at the margins of the world and shows that he was perhaps a naturalist whose work rested in both the medieval and the modern era. Who in the opening passages of his great work does Linné call to as his authorities but the Romans Pliny and Seneca? If Linné had not included *Homo ferus* and *monstrosus*, the chain of being would have been broken. As it was, the break is not between cartography and hierarchy – space versus time – but between the enchanted wonder pervading natural history and the disenchanted world of biology and sociology. The monstrous now became rationalized and new questions were asked regarding these types of mankind. How might one administer the domain of the monstrous? What forms do their governing institutions take? What are their habits, both physical and cultural? And, acknowledging the debt owe to Linné and his concerns, how do they have sex and reproduce? How do their sexual organs distinguish them? It is not so much the institution of a hierarchical system that places Blumenbach and not Linné at the beginning of scientific classification of human variety, but rather the complete rationality of Blumenbach's system. There is no barrier to the administrative and governmental use of the classification of human variety present in Blumenbach's work. With Blumenbach's five-race scheme, the classification of human variety would be placed under the complete authority of science.

Although Blumenbach classified humans on the basis of colour and geographical position, he resisted the temptation to use level of civilization, degree of degeneracy or intelligence and culture as indicators of innate superiority. Instead, Blumenbach argued that any such system is itself essentially arbitrary. 'No general rule can be laid down for determining the distinctness of species, as there is no particular class of characteristics which can serve as a criterion; in each case we must be guided by analogy and probability'.[97] The function of the natural historian is like that of the antiquarian collector: to search through the scattered remnants of past revolutions, piece together a history of nature replete with lacunae and from this weave a natural history. The history of human civilization in fact serves as the measure of natural history; just as nations have experienced revolutions, so too has the Earth been subject to catastrophic changes.

The rationality of Blumenbach's classification was not enough to fix the meaning of human variety. If fixity was the scientific norm and confirmed the biblical chronology of the world, how did one then incorporate the emerging fossil record? Until Darwin's radical recasting, monogenic theory was firmly rooted in the authority of the biblical chronology. Even the fossils in the curio cabinets, gardens and museums were not seen as remains of living creatures, but as aborted creations. There were theories that fossils were simply mineral analogies or aborted discards from the creation of the earth. Accordingly, fossils were not ancient but the result of the same divine force that shaped the animals as it penetrated into and acted on the Earth itself; nor were they abortions, stillbirths or crystallized monsters that the earth could not or would not bring forth to live. Did the myths not speak of humans as moulded from the Earth, and of our returning in death to the dust and ash of the earth?[98] Simply put, the recognition of fossils as the remains of animals and plants no longer living on the Earth – Cuvier would give us the term 'extinction' – did not fit easily with this chronology. One could not put them off to the margins of the map any longer, nor were they monsters or errors. This was Cuvier's intervention into the species question: a system of classification using comparative morphology and a theory of revolution that could simultaneously support a scientific monogenesis and a Christian creation. The catastrophic theory of the Earth could do this only at the expense of allowing into circulation a theory of extinction. An ordered and determined nature was fundamental to establishing the scientific authority of natural history, if not the authority of science in any era. In the fullness of time natural history proved inadequate to the task of interpreting a dynamic nature that was riven by catastrophes and extinctions. Cuvier's attempts to account for change while defending the fixity of species expanded the discursive field of evolution beyond a stale and overdetermined preformism. For a time, Cuvier successfully preserved the authority of religion while saving the appearances of the natural world. Comparative morphology supported the connectedness of species and yet the fossil record contained innumerable extinct species. Because extinction was difficult to reconcile with a view of the fixity of species, Cuvier advanced his theory of revolutions as a solution to this contradiction:

> life on this earth has often been disturbed by dreadful events. Innumerable living creatures have been victims of these catastrophes. Some inhabitants of dry land have seen themselves swallowed up by floods; others living in the ocean deaths when the

bottom of the sea was lifted up suddenly were placed on dry land. Their very races were extinguished forever, leaving behind in the world nothing but some hardly recognizable debris for the natural scientist.[99]

The most pressing contradictions had been exposed but ultimately reconciled by the recognition of geologic stratification, which itself relied upon an understanding of extinction, discontinuity and an earth older than that suggested in the scriptural accounts. The evidence supporting Cuvier's 'theory of the Earth', or what was now called 'geology', indicated that there had been a succession of animal populations. Vast numbers of species had appeared upon the earth at different periods and most had long since disappeared. Species and orders became extinct but new ones appeared. The fossil record strongly suggested to those who took it seriously the creation or the sudden appearance of new forms. The processes of change and extinction are not gradual, but sudden and catastrophic. The evidence for a series of catastrophes could be found in the morphology of animals seen throughout the geo-historical record. For Cuvier the fossils confirmed that a series of great catastrophes had indeed occurred, with each an upheaval of the earth analogous to a social revolution (Table 1.4).[100] Cuvier's monogenic theory reconciled the fossil record, current scientific theories, empirical observation, comparative morphology and biblical authority.[101]

> If there be one thing certain in Geology, it is that the surface of our globe has been subject of a great and sudden catastrophe of which the date cannot go back beyond five or six thousand years; that this catastrophe has overwhelmed the countries previously inhabited by men and by those species of animals with which we are today familiar; that it caused the bed of the previous marine area to dry up and thus to form the land areas now inhabited; that it is since this catastrophe that such few beings as escaped have spread and propagated their kind on the newly uncovered lands; that these countries laid bare by the last catastrophe had been inhabited previously by terrestrial animals if not by man and that therefore an earlier catastrophe had engulfed them beneath its waves. Moreover, to judge by the different orders of animals of which remains have been revealed, there were several of these marine irruptions.[102]

Table 1.4: Cuvier's System of Organic Successions and Catastrophes, compiled from Cuvier, *The Animal Kingdom* (1834).

Organic Succession	Catastrophe (Inundation)
Contemporary species only	Certain
Mastodon, mammoths, etc.	Probable
Paleotherium, Anoplotherium	Probable
First marine animals, marine fish and shells	Certain
Maastricht animal	Probable
Jurassic animals	Probable
No quadrupeds, only fish and shells	?
No fossils	No evidence

Cuvier noted that we are fortunate indeed to have a historical account of the most recent marine irruption: the flood recorded in the Book of Genesis. Unknown to him due to the limited archaeology was the appearance of the narrative of the flood in tales as ancient as Gilgamesh. This would have only given Cuvier greater confidence in his theory. Historical texts and the Earth itself are composed of datable strata which record and represent the history of humans and nature. The book of nature could tell its own history with or without reference to a designer. Although Cuvier obviously preferred the former, he could not have been unaware that the latter was just as likely a possibility. Charles Lyell was to realize this profound reading of the earth in his *Principles of Geology*. Cuvier took the existence of a historical record such as the Genesis story – whether written or oral, scholarly or priestly – as primary evidence supporting the theory of catastrophes and successive creations.[103] But in order to prove the continued existence of this guiding hand, Cuvier was required to reconcile this design with the contrary evidence that history is discontinuous, and the best evidence for discontinuity or extinction was his own work on fossils. Increasingly, the literal interpretations of human origins derived from Genesis fell away under the pressure of geological data.

Cuvier believed that science should be non-religious and not anti-religious. It has been suggested that Cuvier's opposition to evolution was related more to his theory of correlation than to his religious convictions.[104] But this very question speaks to the contradictions between the scientific understanding of human variety and various religious understandings of man. The principle of extinction was intended to support the argument for the fixity of species. For a time natural history had understood fixity as referring literally to an unchanging number of species. But what of the creation of new species? We know that we have lost some but have we gained new ones? What does it mean that fossils demonstrate that today's predominant animals and plants were not so or not found amongst the fossils? Cuvier's work established a theory of the Earth that explained the existence of fossils and their placement gave to the history of nature the history of the fossil record. In preserving fixity and design, natural history was forced to incorporate two key components of enlightened social theory: discontinuity and revolution. These would survive the collapse of natural history and find their place in the emerging sciences of life and society where fixity and design found no place at all.

In Cuvier's succession of geo-historical eras, humans are unique to the present one:

> What is certain is that we are now at least in the middle of the fourth succession of terrestrial animals, that after the age of reptiles, after that of the palaeotheriums, after that of the mammoths, the mastodons and the megatheriums, came the age when the human species, aided by some domestic animals, peacefully dominates and makes the earth fertile, that it is only in the environments formed since this last age, in the

alluvial deposits, in the peat bogs, in the recent concretions, that we find fossil forms of bones which belong entirely to animals known to be living today.

To know the series and to fix the place of humans in the order of nature did not mean that we had a full understanding of the origins of humans. In fact, many of the same questions remained to be worked out.

> When then was the human species? Did this last and best work of the Creator exist somewhere or other? Did those animals which now coexist with human beings and of which there is no trace at all among the fossils, did they accompany human beings? Were the countries where humans lived with them overwhelmed at the same time as those which they inhabit now? When a great flood could have destroyed this previ-ous population, were those countries turned back onto dry land? This is what the study of the fossils does not tell us, and in this discourse we must not turn to other sources ... It is certain that no one has yet found human bones among the fossils. And that is one more proof that the fossil races were not varieties of present animals, because they could not have undergone the influence of human beings ... None of these remains belongs either to the great deposition of the last catastrophe or to those of preceding ages.[105]

The discontinuities in the fossil record made it an incomplete record of the earth's history.[106] Cuvier insisted that other historical knowledges and artefacts, especially biblical exegesis, should be used to supplement the evidence of natural history. Likewise, Linné offered that natural history was the rationalization of our curiosity which was given to us by the creator so that we can become aware of the creator's work and intentions.[107] Far from being our contemporaries, primitives and savages were now to become proof that human variety resulted from multiple creations, and our racial differentiation had its origins in the most recent revolution of the earth.

> The most degraded of the human races, the Negroes, whose shapes most closely approximate the brute animals and whose intelligence has not grown to the point of arriving at a regular government nor the least appearance of coordinated knowl-edge, has preserved no written records or traditions at all. That race cannot therefore enlighten us about what we are looking for, although all their characteristics show us clearly that they escaped the great catastrophe in a place different from the Caucasian and Altaic (Ural) races. They had perhaps been separated from them for a long time when this catastrophe happened.[108]

Cuvier was perhaps one of the most famous exponents of the monogenetic origin of mankind long entrenched in the Judeo-Christian tradition.[109] The popular idea brought into scientific respectability by Linné, and accepted by Cuvier, was that each existing species represents the current aggregate of indi-viduals descended from an original pair. When associated with the theory of the fixity of species, these tendencies – ideology and descent – appeared to Cuvier to necessitate a unitary view of the origins of the human species, and one in which mankind could be arranged using race. These races had thought to be ancient and distinct for such a long period that it now appeared impossible to explain their origins, but it was clear enough that the origins were separate. Having already asserted that climate is an important but not a determining factor in producing variety, Cuvier was reduced to the suggestion that the present human races had escaped in different directions after the last revolution some 5,000 years ago, and had no further contact with each other.[110]

Unlike many others who tended to multiply the number of races,[111] Cuvier held that there are only three: Caucasian, Mongolian and Ethiopian. While each of these could be subdivided into groups according to unique geographical, linguistic and physical characteristics, all were interfertile and therefore constituted a single species. He concluded that

> the Caucasian race, by its monuments, traditions, and religion plainly showed its superiority over the other races. It would be unjust to conclude ... that Cuvier had joined the extreme school of white supremacy. He believed that Negroes, Chinese, and members of any other races were rational and sensitive creatures and that they therefore had as much right as the white to the title of man. Slavery was degrading for both slave and master and must be abolished. Whatever might be the present or past level of relative culture of the many races of mankind, all belonged to the same species, Homo sapiens, and all deserved the dignity and responsibility which were representative of man. The white man's advantage was both obvious and great but it was not to be abused. A beneficent but haughty paternalism characterizes Cuvier's attitude on this subject.[112]

George Stocking reads Cuvier quite differently. Looking closely at Cuvier's memo on the collection of skulls and the determination of racial types, Stocking shows racial classification to be an important concern of Cuvier's. He also finds Cuvier's discussion of the Aryan origins of Egypt to have been important for shaping the mid-nineteenth-century's understanding of race, 'that cruel law which seems to have condemned to eternal inferiority the races of depressed and compressed skulls'.[113]

The thread connecting Linné and Cuvier is the shared imperative to classify nature and know the intention of the designer behind it. There were differences between the two over how this would be accomplished, as might be expected. Cuvier and his followers relied on the concepts of discontinuity and extinction whereas Linné sought the source of the great chain of being in the very order of the table of classification. It was Lamarck who opposed the assumption of fixity in the classification of present species. The work of classification, like the work of empire, is never finished. Where Cuvier had sought to preserve a boundary between man and animality, Lamarck worked instead to preserve the place of man at the head of the great chain of being. 'The arrangement is in the form of a scales or ladder and thus contrasts in principle and appearance with that adopted by

Cuvier'.[114] Despite their differences, both helped to bring an end the study of the great chain and in its place firmly establish the scientific study of man. At the same time both brought forth the means to undermine the authority and uniqueness of man. Lamarck tells us in the preface to his *Philosophie Zoologique* of 1809 that

> in order to fix the principles and establish rules for guidance in study I found myself compelled to consider the organization of the various known animals, to pay attention to the singular differences which it presents in those of each family, each order, and especially each class; to compare the faculties which these animals derive according to its degree of complexity in each race.[115]

Lamarck's quest was for the connecting link between the lowest member of one class and the highest member of the next, and from there, on down the great chain.[116] Despite the allegiances naturalists held either to Cuvier or Lamarck, the central problem of classification united them in a common orientation, animating their imperative to understand and fix the meaning of human variation. 'The stimulus linking [natural history with biology] was therefore that of classification'.[117]

Fundamental to understanding the crises and revolutions of the Enlightenment is locating the new form of science in terms of race and slavery. It does not devalue the achievements of our age of Enlightenment to remember that one of its many ghosts roams about in the garb of the slave-holder. Perhaps it makes the achievement all the more remarkable given its contradictions and terrors. The technology that is an essential aspect of the Enlightenment carries with it a whole set of social relations which can literally be seen in Sade: technologies of pleasure and arrangements of bodies around a machine that both satisfies and produces desire. 'Into hands stained by the murder of spouses and children, sodomy, killing, prostitution, and infamy heaven has placed these riches to reward me for such abominations.' It is in a sense that if all the world is a stage, then the world is also a theatre of cruelty: 'when power was at stake, the rulers have piled up mountains of corpses even in recent centuries'.[118] Is it possible to suggest that the overcoming of race might lead to human emancipation? Such a question could only be answered in the context of the production of knowledge concerning human variety. The power to name, classify and administer was applied to humans as well as to the 'brutes'. Perhaps it would be better to say that the need to administer the 'new' varieties necessitated naming and classification. These classifications came to represent the essence of these new varieties, an essence that could be recognized by anyone now as the everyday truth of racial difference. For a comprehensive classification to remain a possibility, subsequent scientific research developed the concept of race around morphological criteria such as nose and lip shape, hair texture, shape of the facial and lower jawbones. These were considered robust indicators of racial type and useful in the verification of racialist hypotheses regarding the development of human society.

Critical Zoology: The Negro and the Ibis

Lyell's meditations in his *Scientific Journals on the Species Question* show how closely allied the species question was with the many ruminations on human difference. These notebooks chart Lyell's course of research from 1855 until 1861, when he came to accept the transformation of species, a theory which he had previously rejected. Lyell, like his friend Darwin, believed that the fundamental motivation for investigating the species question was to explain the meaning of human variety. Although Lyell did not take the view that races were different species it was not clear on what basis one could separate a variety from a species, and what criteria defined a species as opposed to a race. Lyell thought the line between a variety and a species needed to be clearly delineated and like many others he believed that the study of human difference was the best avenue of approach to the problem. At long last the accumulated resources of natural history would be marshalled to explain the ultimate reasons for the variability of humans. No matter the discussion, from the formation of geological strata to the fixity of species, to the nature of the state, man served as the key referent against which all of nature is measured. The task of natural history was the specific application of 'the physiological principles ... of critical zoology' to the study of human origins and variety.

> All the arguments derived from the structural affinity of the different species of a genus & genera of a family, or order & orders of a class, fall to the ground unless we can apply them to the human family & its different races. It is better to believe with Agassiz & Owen & to follow Linnaeus, that species are all originally distant & not connected by the hidden bond of descent than to grant the hypothesis & not apply the rule to Man.[119]

Without the special place given to man, the alliance of a system of classification with the belief in fixity fell apart. Granting man a place in nature with all other creatures was required by the very structural affinities and comparative morphologies that the system of classification relied upon to create genera and families. The question of whether the study of nature would also be conducted as natural history or as the study of life would be decided by the study of man. Despite his insistence that the world showed the existence of a designer's guiding hand, Lyell knew that the weight of evidence had turned decidedly against the fixity of species. He describes in the journals a conversation in April 1856 at the Philosophy Club between himself, John Stuart Mill, Thomas Huxley, Joseph Hooker and George Busk. Together they agreed 'that the belief in species as permanent, fixed, & invariable, & as comprehending individuals descending from single pairs or protoplasts is growing fainter – no very clear creed to substitute'. But Lyell was well aware that his friend Darwin had already developed the theory of natural selection as just such a 'clear creed'. As we shall discuss below, the usual understanding of this 'struggle for existence' has more to do with Spencer than Darwin, but the basics of natural selection are already here: descent, extinction, adaptation, the generality of selection, etc. The full passage makes clear the importance of human variety for formulating an understanding of speciation:

> Indefinite time & change may, according to Lamarckian views, work such alterations as will end in races, which are as fixed [as] the Negro for example and unalterable for the period of human observation, as are any known species such as the Ibis cited by Cuvier ... If man be modern, & if the Negro & white man have come from one stock, & if such distinct races, if discovered in quadrumana, would have been pronounced species, then new species have been formed since the human pair originated. Vast numbers of living species may be altered forms of extinct fossil forms, & the new climates & geographical position & contemporary fauna & flora may have made the present states of these animals, or plants the only ones which can survive in the great struggle for existence. Just as the black variety of man can alone flourish at Sierra Leone.[120]

Lamarck had suggested in a response to Cuvier that perhaps some species had been more ideally suited to their geographical locale than others, and so had undergone less alteration. That might be the reason why some species appear unchanged over historical time. Just as the ibis was suited to the Nile, so too can the 'black variety' alone survive the tropical climate. This is a general organizing concept in early tropical medicine and commonly cited in popular culture.[121] Lamarck's position was rebuffed by Cuvier in his work on the sacred ibis, by popular lecturers in Egyptology, such as George Gliddon, and by the more serious efforts of the American School; transformism would have to wait until the *Origin of Species* to be finally admitted into scientific respectability. Until Darwin, fixity still had its place in organizing our understanding of human variety. Cuvier argued that the antiquity of the ibis pointed to the fixity of species. Such a highly specialized bird, and one so essential to Egyptian society, had not changed over the span of human history. Cuvier's argument for the fixity based upon the ibis, its depiction in Egyptian murals and as mummified remains would be taken up by Nott and Gliddon to argue for the fixity of the Negro as depicted in some of the same murals.

The ibis and the Negro were seen to have much in common, but they were contradictory representations of fixity as well. The name sacred ibis was given because of its association with Thoth, the god of writing, while the Negro was simply a lesser form of human barely capable of civilization. The lack of any observable change in either the sacred ibis or the Negro during recorded history indicated the antiquity and the fixity of species. If the Negro as the form closest to the beasts had not changed then by implication so too had the Caucasian race remained unchanged as the most beautiful in form and the most civilized in society: the first race with all the significance such a connotation would have later when primitive took on a new meaning. The newly emerging field of Egyptology, which would become popular in the United States thanks to the lecture tour of George Gliddon, provided even more evidence. Egypt's importance was that its texts and artefacts, if read correctly like a phrenologist reading a skull, provided physical objects of study and comparison: mummified ibis remains, funeral goods and a visual and written record going back to the period just after the time of the last great catastrophic revolution or, if you were so inclined, just after the creation. 'Now let us traverse the books of old and their monuments; let us compare what they said of the Ibis and some of the images they traced.' Cuvier had himself examined the remains and depictions of the sacred ibis 'while in ancient Rome' and had examined other specimen brought back by Napoleon's scientists. 'No other animal is so easily observed as the Ibis, as the ancients left excellent descriptions, exact and even colored figures, and the body itself carefully preserved along with its feathers'[122] wrapped tight in triple layers of linen and preservative in a sturdy, life-size urn in the shape of the living bird. In the same passage, Cuvier refers to a wide variety of examples of the ibis which attest to its antiquity – a medal of Adrian, a plinth painting from Herculaneum, the figure of the god Horus, a plinth described by Aelian, a mosaic of Palestrina, etc. – all of which demonstrate that the ibis has remained unchanged over the known history of the earth.

In Cuvier's work the Negro and the ibis do not appear together as they did in Lyell's conversation, but Lyell and his friends understood that the evidence for the fixity of one was also evidence for the fixity of the other human types. Cuvier's purpose had less to do with human variety than it did with defending fixity and design. His 'Mémoire sur l'ibis' is concerned with the problem of distinguishing the sacred ibis from the other birds mentioned by travellers and naturalists as well as clearing up the confusion over the varied names given to the same specimen by Perrault, Brisson, Buffon, Linné and Blumenbach. This fine distinction in Cuvier's favour was of little importance later. The American School would prove the fixity of human variety – and by implication all species – using the same evidence that Cuvier used in regard to the sacred ibis. If the evidence from Egypt proved the immutability of the ibis, the same evidence could and would be used to prove the antiquity and polygenic origins of the races. Race and species were now joined together, infused with complementary meanings which give order to the natural world and everyday social life.

It is a mere coincidence that there were four species of ibis, just as there were four species of human under the Linnean system, but coincidences are convenient especially where they confirm what we already assume to be true. The limiting of human variety to 'four species of different kinds'[123] was the selection of those few important structural characteristics which might easily be compared to each other. The sacred ibis and the Negro became evidence for fixity. Although the Negro would never reach the exalted status of the sacred ibis, it would provide a base for the exalted position of the Caucasian.

The Negro fulfilled the three roles that Lyell described with his Philosophy Club friends. First, as evidence for fixity or at the very least for the 'antiquity of the races'. Second, if the evidence for fixity was falling apart, as Lyell and his friends acknowledged, then the Negro would now take a new position in the order of nature, this time as evidence of the transformation or degeneration of man from its original Caucasian type. Third, the Negro was evidence of the connection between variety and geography. Each type originated, it was believed, in a different locale and appeared ideally suited for it. Whether this was the result of creation and design, or from the degenerating effects of climate, food, etc., was a matter of debate, but this geographic specificity was not in dispute. The Negro provided the definitive evidence that each variety or species is ideally suited to its environment. So the Negro thrives where it is originally found, and it became the self-imposed burden of the European to endure deleterious climates in their civilizing work. A certain weakness and ignorance is acknowledged, though, in admitting the dangers the tropical climate and exposure to savagery pose for the civilized European. Men live in the marginal domains, but no real human can long endure life at the edges of the world or the edges of civilization. After all, as Lyell and his friends already knew, 'the black variety of man can alone flourish at Sierra Leone'. The fixity of the Negro and the ibis served as 'a singular monument of antiquity and decisive evidence of the antiquity of species'.[124] The sacred ibis took its place as a symbol of divinity and fixity, while the Negro was placed in antiquity as a race of slaves. That there were now differences between humans that marked the boundaries between reason and dogma, and war and peace, was agreed upon by all but a few.

> 'It is now clearly proved,' says the illustrious Cuvier, 'yet it is necessary to repeat the truth, because the contrary error is still found in the newest works that neither the Gallas, (who border on Abyssinia,) nor the Bosjesmans, nor any race of Negroes, produced the celebrated people who gave birth to the civilisation of ancient Egypt, and of whom we may say that the whole world has inherited the principles of its laws, sciences, and perhaps also religion. It is easy to prove, that whatever may have been the hue of their skin, they belonged to the same race with ourselves. I have examined in Paris, and in the various collections of Europe, more than fifty heads of mummies, and not one amongst them presented the characters of the Negro or Hottentot.'[125]

2 POLYGENESIS AND THE TYPES OF MANKIND

The American School

Apart from the individual achievements of John Winthrop, Benjamin Franklin, Benjamin Rush and Thomas Jefferson, the first real triumph of science in the New World was the formulation of polygenic theories of human origins by what became known as the American School.[1] James Dekay proclaimed that as America had rid itself of 'our colonial situation, [and] the embarrassments arising from our exposed and peculiar position ... Those interested in Natural History were before then too widely scattered over this extensive country to allow of that familiar interchange of opinions which necessarily elicits further inquiries and discoveries'. This had changed by the end of the War of 1812. Major expeditions to map out the continent and 'enlarge the boundaries of Natural science' were undertaken. Improvements in communication and the growth of the population created the basis for natural history to become an area of serious inquiry in the United States. Already in 1826 Dekay could speak with nationalistic pride of the accomplishments of American naturalists and the

> progress made in Mineralogy, Geology, Botany, and Zoology ... a spirit of inquiry has been awakened. The forest, and the mountain, and the morass have been carefully explored. The various forms and products of the animal, vegetable, and mineral kingdoms have been carefully and, in many instances, successfully investigated. A proper feeling of nationality has been widely diffused among our naturalists, a feeling which has impelled them to study and examine for themselves, instead of blindly using the eyes of foreign naturalists, or bowing implicitly to the decisions of a foreign bar of criticism.[2]

This nationalism would animate the scientific ideology of the polygenic theory as well: Samuel J. Morton, George Gliddon and Josiah Nott, Louis Agassiz and George Squire became prominent among the many natural historians, physicians and anthropologists whose work gave polygenism the cloak of scientific respectability. Accordingly, they were perhaps the first American scientists to be fully recognized by their European counterparts. It is sometimes argued that the mantle of respectability was bestowed upon them through their association with Louis Agassiz, the great anti-Darwinist, but Agassiz came late to the monogenic/polygenic debate and so took on the role of proponent rather than that of originator. The American School had already laid a scientific basis for the polygenic origin of humans and the resulting classification of racial types. By 1850, it had accomplished one of its primary goals – the critique or outright overturning of the biblical chronology of the history of the earth and its inhabitants – and it was one problematic for Agassiz, the zealous Swiss émigré. Gossett notes that the polygenic theory waned in European scientific circles after Cuvier's debate with Saint-Hilaire.[3] But in America, naturalists drew on Cuvier's work on geological upheavals and the fixity of species while simultaneously dispensing with any commitment to the biblical chronology.

But it would be a gross exaggeration to see the American School as a sort of Cuvier without a deity. Although, Nott and Gliddon honoured Cuvier by having his likeness represent the 'Caucasian-type' in their book (Figure 2.1), nonetheless, their works and others of the American School were read by their fellow citizens with an 'emotion of national pride' that was felt even by their monogenist opponents.[4] Freed from dogma, the American School was able to begin what mattered most to its adherents: 'the free scientific inquiry' into the nature of human variety.

The contest between the monogenic and polygenic theorists was not between supporters of two obviously erroneous theories, but between two powerful resolutions of the meaning of human variety and, with it, of the more general species question. It would be simplistic to think that the monogenic/polygenic debate was simply one between pro- and anti-slavery advocates that was temporarily displaced into the realm of science, and then banished forever by Darwin. The complexities of this debate went to the core of scientific inquiry. Both sides – if indeed it is possible to reduce the dispute to only two sides – advanced views that Darwin later accepted or repudiated. Supporters of slavery could be found on either side of the issue, as could abolitionists. Audubon's co-author, the monogenist Rev. John Bachman of Charleston, was a supporter of slavery while George Squire, polygenist and founder of the New York Anthropological Society, opposed it.[5]

The polygenic theory contributed to the establishment, institutionalization and authority of American scientific work. The theory attracted so many because it served three interconnected functions which have passed down to us in our common-sense notions of scientific inquiry. The first aspect of polygenism was its repudiation of the biblical chronology of creation and design. This chronology had of course been the subject of long dispute. Religious scholars had set various dates to the creation of the world, settling on a date of around 6,000 years ago if the genealogies in Genesis are taken as true, and dating the flood at about 3,000 years ago. Cuvier himself had not upset this scheme with his introduction of catastrophic revolutions of the Earth. Until the polygenists, the debate over the

Figure 2.1: Illustration to accompany Agassiz's 'Sketch' in Nott and Gliddon, *Types of Mankind* (1855), detail of foldout at p. lxxvi. The portrait of the European type is of Cuvier. From the author's collection, 8th edn (1860).

biblical account was not centred on whether it was a true account of the history of the earth, but instead on which was the most accurate reading of the biblical text, with some famously calculating out the Mosaic genealogies to establish the moment of creation to the day and time. The polygenic theorists based their critique of theology on evidence drawn from diverse work in anatomy, philology and archaeology. Despite their defence of fixity and polygenism, the American School emerged from the same scientific backdrop as Darwinism, but from a very different political one. Darwin 'appropriated the time scale of the geologists. But by [the polygenists'] incessant hammering at the biblical chronology they did help to prepare the public mind for the Darwinian chronology'.[6]

The discrediting of the biblical chronology was constitutive of the second aspect of polygenic theory: the principle of free inquiry and the separation of religious from scientific/philosophical knowledge. Free inquiry was not simply the pursuit of knowledge for its own sake, but the production of a knowledge which orders the world. The polygenists display in bold relief the two sides of Enlightenment: the freeing of rationality as well as the use of Reason to dominate both nature and humans by other humans. Often slavery and the Nazi exterminations are seen as distinct from each other, and the most horrific of individuals boldly proclaim which one was worse than the other, which one was more oppressive. But slavery and the death camps are bookends to the Enlightenment, to that tendency in Enlightenment to use the same rationality which had promised humanity so much to dominate both humans and nature. Polygenism cuts a wide arc from American slavery to Nazi Germany, and down to the racialist thinking of our own everyday life that still resists whatever progress has been made.

The third aspect of polygenism is aligned with this double-sidedness of Enlightenment. With Enlightenment, scientific knowledge had come to be accepted as a means to better the individual in particular and society in general. Scientific knowledge was now producing a discourse on life and society that could be widely deployed to analyse and resolve even the most difficult of social problems. All of the disciplines that we think of today as having long and venerable histories emerged quite recently and were required to prove their status as sciences through the ordering and improvement of life and society, in other words, as governmental knowledge.

These three aspects of polygenism are given in order of their emergence in the debate over the species question. It might be noted here that the career of polygenism was a comparatively short one as accepted science, but it has a long one as a scientific ideology.[7] The period when the American School brought forth polygenism as the legitimate science of man and during which it dominated social and naturalist discourse can be most conveniently placed in the years between the 1839 publication of Samuel G. Morton's *Crania Americana* and when the publication of Darwin's *Origin of Species* transformed it into a wretched and forgotten knowledge. The extent of Darwin's influence rightly shades all the modern sciences of life and society: so great is his stature that all other rival traditions have been reduced to the level of pre-science or ideology. Darwin presented a thoroughly modern theory of the origin of species precisely because the rival theories of his day – especially of the American School – were thoroughly modern as well.

To their credit, the American School and its allies helped formulate the possibility for a real history of the world. They sought to replace religious dogma with rational evidence drawn from observation, experimentation and the most current work in anthropology, philology, medicine, anatomy and archaeology. They even turned the 6,000-year period given by Genesis into an argument for the polygenic origins of human variety.

Josiah Nott's 1846 article on 'The Unity of the Human Race' contains the essential lines of attack to be used in the American School's critique of religious authority and the biblical chronology. Nott feigns the desire for a resolution of scripture with recent scientific advances, but from the start he clearly indicates his desire to lay waste to the biblical chronology and with it the authority of scripture in scientific work. Far from being a challenge to orthodoxy, Nott says that polygenesis is the only view that strengthens the belief in a creator.

> A denial of the Unity of the Human Race, so far from infringing on the veracity of Scripture, will, I am satisfied, become one of the most solid grounds of its defence; and I venture to predict that, at no very distant day, this opinion will be maintained by a large body of divines.

He asks, which is more important, the unity of the races or the entire history and plan of creation? Although Nott believed in neither, he knew that by rhetorically splitting them off from each other, he could undermine both. From the outset, Nott has his religious opponents in a rhetorical bind. If they accept the chronology derived from the 'Mosaic account', then they are obliged to accept the plurality of races for the simple fact that there has not been enough time for climate, diet, etc. to cause such great variation in humans. 'There is no rational chronology, yet fixed, which will allow time for this wide-spread and diversified population from a single pair, and the facts can not be explained, without doing violence to the Mosaic account.'[8] If Nott's opponents rejected the traditional analysis that put the creation of the world only 4,000–6,000 years ago, they had also to reject the literal authority of the scripture. It was precisely this authority that underpinned the monogenism of the abolitionists. Sure that his opponents could not account for both human variety and the age of the earth without fatally contradicting themselves, Nott remarks that some would attempt to get out of this bind by claiming that the flood was not universal, but he says that this too does violence to the 'Mosaic account'. No one could reasonably doubt from the language of the biblical account that the flood was anything less than a universal catastrophe.

In a footnote, Nott mentions a recently received letter from George Gliddon detailing his newly discovered 'Chinese chronology' which pushed the origins of China, Egypt and India 'within two centuries of the Septuagint date of the flood'. This agreed with views he had himself put forward in an earlier address on the antiquity of the human races:

> I have shown in my previous communication, that the physical characteristics of the races, as seen at present in the Egyptians, (though mixed) the Chinese, Hindoos, Negroes, etc., existed four thousand years ago; and I shall show, before I close, that if physical causes can change a Caucasian into a Negro, it requires a series of [world] ages for its accomplishment.[9]

Given that belief in intelligent design was even more fashionable in Nott's day than our own, even to allow for the possibility of transformation – albeit as a rhetorical ruse – was quite sophisticated and rightly suggests that the species question was already being marked out as Darwin was returning from the voyage of the *Beagle*. The steady accumulation of research around the species question produced many objects of study, some of which have passed out of accepted knowledge and some of which remain. So too have institutions and disciplines such as sociology, biology, geography or ecology emerged from the cacophony of scientific statements about species, variation and civilization. Before these disciplines, the question of change and the question of human variety had already been the concerns of natural history, philology and political economy. These disciplines are set apart from those we have today by the context of slavery, supremacy and institutional relationship to the government's administration of the nation's populations. Neither religious dogma nor political compromise could resolve the social contradictions of slavery and supremacy, nor could they repress the will to freedom manifested by slaves themselves. These problems had become the domain of scientific and medical inquiry. This is how the species question and human diversity came to animate both natural history and the emerging sciences of life. After Darwin's answer to the vexed question of human variety, sociology and biology took up the task of explaining the meaning of this new approach to diversity, heredity and social reform by those resolute in doing the social work of the new era.

Nott continues his essay by referring to his earlier well-received lecture series where he began

> with the proposition that there is a genus Man, comprising two or more species. In the first place, it is all important to determine what is meant by the term species, – whether there are any criteria which will enable us to draw distinct lines between the species of different genera – and whether that genus homo, with its species, does not

stand upon the same ground as others. Here, it is evident, turns one great difficulty of the discussion.[10]

It was a difficulty which the polygenists were determined to resolve. If there are, as naturalists agree, different species of lemur, are not the things that distinguish them similar to the distinctions found in the variety of humans? Nott understood that the answers to this question had grave scientific, social and religious implications. Characteristically, he turned to his own authorities – Blumenbach, Virey, Prichard – for the resolution of this 'great difficulty'. He acknowledged, though, that these same scientific authorities were ambiguous on the species question. Blumenbach, author of the now common classification,

> whom there can be no higher on a point of this kind, tells us, that 'No general rule can be laid down for determining the distinctness of the species, as there is no particular class of characteristics which can serve as a criterion; in each case we must be guided by analogy and probability'.[11]

But of course Cuvier had already suggested key criteria: morphology and interfertility. And so Nott turned to Virey, who had maintained what was for him the correct position: 'Certainly if Naturalists were to see two insects, or two quadrupeds so consistently different in exterior forms and permanent colors as the White man and Negro, in spite of their interfertility when crossed, they would not hesitate to establish two distinct species'.[12] Here again human variety gave meaning to the general variation of nature. It was not simply a question of whether species exist, but – and more critically for understanding nature, society and the relation between the two – a question of their material differences and historical mutualism. The definition of a species was and still is difficult to establish, but the definition of the human species would establish the meaning for all of nature. The explanation for the variety of humans would provide the answer to the species question.

While few could doubt from the accumulated evidence of science and the senses that variation existed, this realization had not yet found its appropriate scientific expression. Nott turned from Virey to Prichard's *Physical History of Mankind*, where he found much to support his view on fixity and human variation. 'It appears ... that in mankind, as in some other races, particular varieties are adapted by constitution, and physical peculiarities, to peculiar local situations.'[13] Variation was not proof of the effects of climate and food. On the contrary, the extent of variation and the antiquity of the races despite so many migrations and forced deportations proved that climate was only a minor factor in variation. Varieties were 'peculiar' to their locales and the antiquity of the races proved that climate and general environmental differences could not fully account for the extent of variation, even less so if one is committed to the biblical chronology.

It must be true, Nott concluded, that the species of a given locale were created only for that 'peculiar local situation', and so each species and its habitat is fixed in time and space. This 'realm' can only be changed by a catastrophic revolution of the Earth. The great antiquity of the races established the fixity of species just as the fixity of species explained the great antiquity of the races.

> Admitting great changes in man from climate, etc. we have little ground for believing that one thousand years could transform the Anglo-Saxons into Negroes, or Negroes into Ourangs. How could a reasonable man believe that any thing short of a miracle, could, in the Temperate parts of Australia or America, change the white race into Australians or Indians. Observation and history alone can settle these points, and we have no record of such changes having occurred, or being now in progress ... Lyell and others tell us, that very few generations are sufficient to effect all which can be effected in animals, and we are informed, also, that these changes are effected with great certainty and uniformity; but history, from the time of Herodotus to the present, affords no positive evidence of these changes in man so 'plainly taught by analogy'.[14]

For those natural historians working towards a scientific understanding of human variety, the antiquity of the various forms made their classifications possible. Just as important, this understanding made the conditions for a new scientific history of nature to be told. Where historians lacked a text or an archive, the bodies of the living and the dead could fill in the lacuna. Bodies may not be historical texts, but they do have their histories individually and as representatives of groups and varieties. An adequate classification of human varieties held the promise of providing the foundation of the 'genealogy of nations' that was at the centre of scientific interest.

> The physiological study of the human races may thus often become a useful auxiliary of history, as M. Edwards has so well proven by examples; and sometimes even it may, when history is silent on the origin of a colony, supply the deficiency, unite the interrupted thread of tradition, and by reading the past in the present, reestablish the genealogy of nations.[15]

Given the available evidence, Nott admitted that the biblical account and the effects of climate were insufficient to explain human variety and so, he wrote, one is left with the task of putting forth a scientific and rational theory of separate origins and a plurality of species. He declared that the world is either not old enough to have allowed the climate enough time to alter the original stock into separate races – and thus the races must be fixed by design and at best the 'Adamic account' refers only to the creation of the Caucasian race – or the earth must be much older than the 6,000 years of the accepted biblical chronology. The great age of the earth was supported by the new fields of Egyptology and archaeology, which were emboldened by new discoveries and novel interpretations of hieroglyphics. Did not the ancient inscriptions depict the same types of mankind as exist today, Nott asked, just as Cuvier had shown that they depict the same sacred ibis? Was not this proof of the antiquity of the races and the fixity of species? This is a point where Nott's account intersects quite nicely with the recent work of Livingstone on the concept of 'pre-Adamic' races. His concern is with understanding the polygenic theory from the perspective of the speculations on pre-Adamic humans which often sought to reconcile polygenic theories with religion. Although some attempted to avoid this controversial use of the polygenic theory, most of the naturalists who were drawn to polygenism were not interested in reconciling religion with their new science. Religion had nothing to fear so long as it accepted the separate origins of the races. Livingstone gives special attention to the social and political implications of the survival of polygenism in the theological concept of the pre-Adamic races that continues to animate supremacist ideologies. In this way, our approaches to the importance of the polygenic theory complement each other.[16]

In the realm of natural history the only remaining scientific objection to the polygenist position was the interfertility of these supposedly separate species. The same antiquity which seemed to speak of the fixity of the races was now an obstacle, because the inter-fertility of the races could only be as ancient as the races themselves. Cuvier had defined a species by the ability to produce viable offspring and this view held increasing sway. Founded on essential differences between the master and slave, modern slavery had produced unions and the body of the mulatto gave ample evidence of the interfertility of Caucasians and Negroes. Needless to say that confronted with this problem the question was quickly shifted away from interfertility and focused instead on the viability of the hybrid's descendants. Hybridity was assumed to result in sterility, if not in the first generation then ultimately within the next few generations. Was the mulatto, as the hybrid *par excellence*, viable? What of the octoroon or the quadroon? Jefferson had speculated in the *Notes on the State of Virginia* that the mulatto was probably of a stronger constitution than either of its parents and this long before his encounters with Sally Hemmings. In contrast, he believed that the pure-bred African slave was incapable of living in a civilized society without the supervision of a master. Jefferson saw the continued presence of Africans in America ultimately producing ceaseless civil strife. Slavery had already led to the moral degeneration of Africans from their contact with civilized vices. Enslaved and transported from their native realm, they easily fell prey to the vices that they now openly displayed in society. This behaviour had been erroneously believed to be common in their native Africa but was instead the product of contact with civilization. The nature of the Indian had also been altered by these same vices, but in the case of the Africans, their nature had in addition been stamped by the degradations of slavery. Those who had been moulded by slavery could only be set right again by repatriation to Africa and Jefferson was sure that when the problem of slavery was solved by a future generation, the necessity of repatriation would be clear. On the other hand, there was no need to repatriate mulattoes, as they would in all probability make significant contributions to the Republic.[17] The effects of amalgamation on the mulatto's moral and physical nature were take to be 'perhaps the most important of all points connected with the Natural History of man'.[18] The questions surrounding hybrids and the effects of amalgamation did not develop after the scientific ideology of polygenism emerged, these questions were there from the very beginning. It was around the subject of hybrids that American natural history could arrange the study of society and the problems of government. Henceforth the American School abandoned most of its Jeffersonian tendencies.

Nott's method was consistent with Saint-Hilaire's and Cuvier's use of texts where no physical evidence existed. He maintained that the natural history of man necessitated a new science which would grant the existence of essential differences between the varieties of humans and put race at the centre of the study of man.[19] Only then might this new science of nature 'cut loose the history of mankind from the Bible and place each upon its separate foundation where it may remain without collusion or molestation'.[20] Indeed, the biblical chronology could be abandoned when the fixity of species could be assumed. Without the collusion or molestation of religion natural history could demonstrate that the mulatto is governed by a morbid 'law of hybridity'. The monogenic theory rested upon religious authority and this authority gave moral and material weight to the anti-slavery movement. During the monogenic/polygenic debate, each side would claim a basis in reason, but with the success of polygenism, abolitionism lost its basis in reason and relied instead on its biblical foundation.[21]

Nott derived many of his views from his medical practice in Mobile, Alabama.[22] From his observations, he offered up for his readers six fundamental characteristics of the mulatto which he thought 'corroborated very strongly this idea' that Caucasians and Negroes are 'distinct species'. In his experience as a physician, Nott said he found mulattoes to be 'intermediate ... between the black and whites' in intelligence. In terms of physical endurance and resistance to pain, the mulatto was inferior to either of its 'parent races' and was in general unable to suffer long the 'fatigue and hardships' of physical labour. Mulatto women, Nott argued, were 'particularly delicate' and tended to suffer from a number of chronic diseases. This last point was the foundation for his next: any congenital or long-term disease was a strong indication of consanguineous decay. This combination of a lack of physical endurance and an overabundance of chronic disease led to Nott's fourth observation: 'that the [mulatto] women are bad breeders and bad nurses'. Many mulatto women were either sterile, he observed, or endured a greater frequency of spontaneous abortions. Nott found mulattoes of both sexes to be less prolific if they intermarried, and so must 'cross' with 'one of the parent stocks' to remain vigorous. This final point is at first glance somewhat curious because it seems much more specific than the others. It is here that Nott's work on yellow fever came to contribute to his answer to the question of hybrid vigour. Negroes and mulattoes 'are exempt in a surprising degree from yellow fever', a characteristic Nott derived from two sources: inherited predispositions to chronic disease and a resistance to diseases of 'chance and environment'. The weak constitution of the mulatto was the result of climate. Like the European, the mulatto was predisposed to degenerate in tropical climes. The mulatto was more resistant to yellow fever than the European and so Nott argued that on the basis of relative susceptibility to disease the mulatto was ultimately closest in constitution to the Negro than to the European. One might perhaps assume that this greater resistance would be a sign of great vigour, but because it indicated to Nott a closeness to the Negro, the vigour it might bestow was not enough. Through these 'bad breeders and bad nurses', one drop of blood weakened the entire line of descent and, of course, it was as usual all the mother's fault.[23] One is left to wonder if Nott ever reflected on his own seeming good fortune not to have degenerated or worse in the humid heat of nineteenth-century Mobile.

The results of the 1840 Census certainly provided ample scientific and moral evidence of this supposed stain. Although the evidence satisfied both the advocates of climate-induced variation and the advocates of fixity, Nott said that he was prompted to turn his attention to the species question by the results of this census. Stanton wrote that Nott was motivated to take up his work on racial classification after reading the results of the census as reported in an 1842 article in the *Boston Medical and Surgical Journal*.[24] The author identified himself only as 'Philanthropist' and worried that all of the attempts to improve the lot of the Negro would now have to be changed due to these new facts. Now an almost forgotten moment in the progress of the social sciences, the 1840 Census nevertheless had an important place in the politics of the period leading up to the Civil War. For the first time, manufacturing and trade data was collected. The country was already sharply divided between a largely agricultural economy and an increasingly powerful and developed industrial sector. Manufactures and trades were added in order to give government some information about the different social formations of the North and the South and to allow policymakers to compare them. The question of slavery was itself a question of what type of economy America would have, and about the virtues and vices of each. It came naturally that one of the only American epic poems – *The Hireling and the Slave* – was composed to contrast the conditions of the slave, who was depicted as well cared for and increasingly civilized by regular labour, with those of the northern mill worker, whose poverty and degeneracy were depicted as much worse. It was, as one might expect, a poem written by a Charlestonian, and a best-selling response to abolitionism and to Stowe's *Uncle Tom's Cabin*. It's author, William

Grayson, was a southern moderate who largely opposed succession. *The Hireling and the Slave*, Parrington says, was 'vigorous propaganda' proving that '[i]f the South was attacked, it was not without weapons to defend itself'.[25]

The 1840 Census brought the debate over the scientific classification of human variety into the centre of public policy debates, and the fact that these classifications had real political implications was not lost on anyone.[26] The 1840 Census is just one of many events where we can see just how social scientists and others have often, in the words of a key player in the drama, Edward Jarvis, 'unconsciously sent forth error'.[27] That the census was a political production as well as a scientific one was not lost on those in the nineteenth century, just as it was not missed by those arguing for the inclusion of a multiracial category for the 2000 Census. The politics of the proposal are easily found in the administration's rejection of the change. The 1840 Census marked shifts in American society and social thought, and was only the second time that the insane, idiots, morons, the blind, the deaf and dumb were counted, including whether these last were in the public or private charge.[28] The racial categories used in the 1840 Census were generally those already established by Blumenbach.

The census takers fanned out over the country. Their findings took many months of painstaking work to compile and finally publish for Congress. Edward Jarvis, a physician from Concord, Massachusetts who in later life would later become one of the founders and president of the American Statistical Association, began his association with the census by looking for data of interest to the new field of vital statistics. Jarvis and the director of the 1850–70 censuses, John DeBow – the editor of *DeBow's Review* – were instrumental figures in the development of vital statistics and social studies, as well as, in DeBow's case, the popularization of scientific knowledge about human variety.

Jarvis developed what he described as a great hatred of slavery during the six years he practised medicine in Louisville, Kentucky. Here his story takes a twist, for Stanton says that Jarvis made a startling discovery, whereas Anderson says that Jarvis was only brought in later to check for errors. Yes, he was, but only after he was so shocked to find that he had 'unconsciously sent forth error' and sought desperately to correct it.[29] In analysing general trends in the census reports, Jarvis found that the rate of insanity among whites was statistically the same in both the South and the North, while Negroes showed a marked disposition towards insanity that appeared to intensify with increased latitude. The further north Negroes lived, the more likely they were to be counted amongst the insane, idiots, morons, the blind, and the deaf and dumb. The differences were strikingly wide: 1 in 162 Negroes in the North was listed as insane, while the statistic for southern Negroes was only 1 in 1,558.

In Maine, every 14th Negro was insane or an idiot, in New Hampshire, every 28th, in Mass every 43rd, in New Jersey, every 297th. Just over the Mason–Dixon line in Delaware, it dropped even more to 1 in every 600. The Mason–Dixon line, it was said, marked the line between sanity and madness.[30]

Although some tabulation errors made Jarvis suspicious from the beginning, he nonetheless concluded that the data indicated that in the end 'slavery must have a wonderful influence upon the development of moral faculties and the intellectual powers' of the Negro. The evidence clearly showed, he said, that the Negro could not endure the 'liabilities and danger of active self-direction'. Freedom in the North was a 'false position' that pushed Negroes to the brink of and often into the abyss of madness. Social statistics had proven the correctness of the South's institutions of slavery.[31]

Jarvis though, was from Concord. He had schooled alongside Emerson and others, and had not forgotten the horrors of slavery that he had witnessed. Confident though he was in his findings, he decided that further investigation was in order given the extraordinary nature of the results. Since his analysis was published in serial form, we have the interesting case that on pages 116–21, Jarvis propounded his results regarding freedom and madness, and a few months later, on page 281, he began to raise numerous questions about those same findings. A further three months of research followed, during which Jarvis examined the reports from every city in the North and found that several 'reported to have as many colored lunatics as people'. He was sure that these errors crept in during the complex computation of the raw data from thousands of reports. Jarvis had a somewhat naive but very modern conception of social science. He was, he said, 'disappointed and mortified [that he had] unconsciously sent forth error', but was sure that now the census would be corrected.[32]

While Jarvis was trying to correct his errors, more popular works – which were often far more serious than we are accustomed to today – took up the call. Using the authority of these errors the *Merchant's Magazine* proclaimed that the census proved not only that freedom but also the 'northern winter ... affects the cerebral organs of the Negro race' and that slavery held down the rate of insanity among Negroes. The *Southern Literary Messenger* spoke of the misplaced philanthropy of abolitionists, who would condemn Negroes to madness and whites to a countryside where 'maniacs and felons meet the traveler at every crossroad'.[33]

Three months later, Jarvis published an article on the census errors in the *Boston Medical and Surgical Journal*, forerunner of the *New England Journal of Medicine*, where he was a frequent contributor and where he would in the future often turn to publicize his refutation of his early results. Jarvis argued that because there were so many errors in the census, nothing short of a 'refutation ... coming forth with all the authority of the national government' and 'corrected by the Department of State' could resolve the issue. Already, he said, 'throughout the civilized world' the word had spread that he had proved that slaves were suited by nature for slavery in the sub-tropics, and that freedom left them with 'with insane delusions ... or crushed in idiocy'. Jarvis ridiculed these findings, noting that cases of deafness and dumbness among Negroes were recorded in towns with no Negroes, and that in other towns the entire Negro population was noted as afflicted with problems of blindness, insanity, deafness and pauperism. It was, he argued, necessary to redo the census for 'the honor of our country, for medical science, and for truth'.[34]

Jarvis was ignored. The politicians and scientists of the time had already heard what they were predisposed to hear. In 1844, while too closely admiring a new cannon, Secretary of State Abel Parker Upshur was amongst those killed in the explosion aboard the *USS Princeton*. Senator John C. Calhoun of South Carolina was then appointed Secretary of State and he came into the Cabinet with two problems. First, the claims by Jarvis and others regarding errors in the census, and second, the status of Texas as a slave territory. Great Britain was pressing the United States for Texas to remain independent and slave-free. Calhoun would have none of it and found it fortuitous that the solution to either problem could lead to the solution of the other. The annexation of Texas was a done deal, he held, and moreover the 1840 Census had shown conclusively that '[i]f the experience of more than half a century' was a guide, then abolition would be neither humane nor wise, because 'the census and other authentic documents' proved 'that the condition of the African invariably sank into vice and pauperism, accompanied by the bodily and mental afflictions incident thereto – deafness, blindness, insanity, and idiocy'. In the states that had retained the ancient relation of master and slave, Negroes had 'improved greatly in every respect – in numbers, comfort, intelligence, and morals'. Calhoun argued that slavery was necessary for the preservation and refinement of the Negro race, and the annexation of Texas was necessary to preserve slavery.[35] He had already said in an 1837 Senate speech that in the southern United States,

> [n]ever before has the black race of Central Africa, from the dawn of history to the present day, attained a condition so civilized and so improved, not only physically, but morally and intellectually.
>
> In the meantime, the white or European race, has not degenerated. It has kept pace with its brethren in other sections of the Union where slavery does not exist. It is odious to make comparison; but I appeal to all sides whether the South is not equal in virtue, intelligence, patriotism, courage, disinterestedness, and all the high qualities which adorn our nature.
>
> But I take higher ground. I hold that in the present state of civilization, where two races of different origin, and distinguished by color, and other physical differences, as well as intellectual, are brought together, the relation now existing in the slaveholding States between the two, is, instead of an evil, a good – a positive good

Calhoun's purpose is to compare Europeans to Negroes and not southerners to northerners because there is no reason to do so. What Calhoun wanted to do was defend the institution of slavery. There was really nothing unusual about southern slavery, Calhoun maintained, if one looks back over the entire course of history. All civilizations rested on the exploitation of the many by the few. Slavery is just one method of coercing labour from the many, falling on a spectrum

> from the brute force and gross superstition of ancient times, to the subtle and artful fiscal contrivances of modern. I might well challenge a comparison between them and the more direct, simple, and patriarchal mode by which the labor of the African race is, among us, commanded by the European. I may say with truth, that in few countries so much is left to the share of the laborer, and so little exacted from him, or where there is more kind attention paid to him in sickness or infirmities of age. Compare his condition with the tenants of the poor houses in the more civilized portions of Europe – look at the sick, and the old and infirm slave, on one hand, in the midst of his family and friends, under the kind superintending care of his master and mistress, and compare it with the forlorn and wretched condition of the pauper in the poorhouse.[36]

Calhoun thought the matter done for at least the time being, but presently the French government stepped in to raise anew the issue of the expansion of the slave territories. Needing even more scientific backing this time, Calhoun turned to the former American consul in Egypt, George Gliddon, for help with understanding the current scientific consensus. Gliddon had recently made quite a name for himself. He had just finished a lecture tour which helped invent the mid-century craze for Egyptology in the United States. In later years Gliddon would convince the US Army to create a camel corps to patrol the south-west deserts, grow bitter that he was not chosen to head the project as he was not actually in the military, and die of fever in a failed venture to build a canal across the isthmus. Gliddon referred Calhoun to the work of Samuel George Morton of Philadelphia, who had already published works on Native American and Egyptian skulls and who had rejected phrenology in favour of the scientific study of craniology and physical anthropology. Gliddon sent Calhoun both of Morton's crania books, an article where Morton suggests the multiple origins of the races, and some of his own writings. Gliddon had himself contributed to Morton's collection of Egyptian crania, sometimes through legitimate means and at other times even paying for graves to be robbed and their contents shipped to Morton. Gliddon, with an exuberance that the shy Morton studiously avoided, told Calhoun that Morton's work proved that the biblical story of the origins of humans was false, that races are distinct species of human, and that Negroes have 'ever been Servants and slaves, always distinct from, and subject to, the Caucasian, in the remotest times'. Taken by Calhoun's interest, Gliddon excitedly informed him that in America there existed a group of naturalists, historians and physicians with 'any amount of facts at our disposal to support and confirm all these doctrines', should Calhoun 'desire the solution to any ethnographical problem, in respect to African-subjects'.[37]

Calhoun's adversary in the Senate was the former President and opponent of slavery John Q. Adams, who took up the call by Jarvis to alter the 1840 results. Ultimately, Adams's attempt to have the census invalidated failed as a result of Calhoun's more skilful parliamentary manoeuvres. The effort to have the findings re-examined was approved though further undermined by Calhoun retaining, as Secretary of State, the power to appoint the supervisors of the census, and he promptly appointed the very same persons who had served as supervisors in 1840. It should come as no surprise that their review found that there were no substantial errors and officially the census stands as a legitimate government document. In the meantime, Jarvis continued his association with the census while trying to correct his errors. He helped plan the 1850, 1860 and 1870 censuses, and became an early leader in creating the American Statistical Association to support his efforts to revise the 1840 Census. By 1850 DeBow would become both head of the census and leader of the Statistical Association. Jarvis was steadfast throughout the years in pursuing changes to the 1840 Census. Towards the end of his life, he wrote a three-part essay in the *Atlantic Monthly* titled 'The Increase of Human Life' where he argued that the ability to collect and preserve records and vital statistics had historically been the difference between civilization and barbarity. The expansion of administrative techniques, the professionalization of public health and social reform provided a 'sanitary history of the world' that argued against those who claimed that 'these are the days of degeneracy'.

> Those whose years are weakened and whose lives are shortened are no more fixed by any law of nature in their present low vital condition than were their fathers, ages ago. There is nothing more in their personal organization or social relation, to prevent their improvement, than there was with people similarly circumstanced in times past. These have the same power as their fathers, and more opportunity, to amend their condition and their habits, to increase their strength and diminish their ailments, and prolong their days on earth. And they will do so; the present and the coming generations will go on in this good way; each will make some progress, and to each successively will be given a larger, richer, and longer life.[38]

There might have been much debate over the results of the 1840 Census, but there was little debate over the racial classifications upon which it rested. The debate over the origins of human variety was acknowledged as over with the acceptance of the polygenic theory. Race explained and marked the meaning of variety: 1. the races are evidence of the separate origins of humans; 2. the races are different species of human; 3. Negroes were the most recently created and therefore most primal and Europeans the most advanced in terms of culture and physique. Europeans lost their place as the originating 'primitive' species in favour of being the most civilized.

The American preoccupation with the monogenesis/polygenesis dispute contributed much through 'the incorporation into European writings of a large body of observations from the New World relating to Indians, Negroes, and animal life'.[39] The names of Wells, Morton, Nott, Gliddon, Bachman, Audubon and later Agassiz were those of central figures in the most intensive and important scientific and social question of the day. These were people who sincerely believed in free thought and scientific understanding liberated from the constraints of a dull Christian dogma. They viewed the future as the domain of reason and science. Theirs was less a contest between the science of the New and Old Worlds than contrasting deployments of a particular scientific ideology in the different centres of scientific knowledge. Edinburgh, Philadelphia and Charleston were equally important nodes for the study of human diversity. Prichard and Morton both studied in Edinburgh and the monogenist John Bachman of Charleston studied intensely the physical characteristics of Chinese and African sailors he observed in Liverpool during his European tour.

Samuel Morton was one of the few naturalists on either side of the Atlantic who Josiah Nott respected. Morton's rise in Philadelphia to his place as the most prominent American scientist has been best discussed by Stanton and by Gould, who relied greatly on Stanton's account. Morton gained a worldwide reputation for his collection of over 900 human skulls and 600 other animal crania.[40] This collection provided the empirical basis for his research into the defining characteristics for the classification of human variety.[41] Morton's position as the president of the Natural History Society and membership in the American Philosophical Society came with his increasing prominence. It was Morton who was entrusted with examining the crania collected by the United States Exploring Expedition of 1838–42, though he died without completing his work. E. G. Squire wrote at Morton's death that

> As Americans we may take pride in the reflection that an American physician, with the aid of a few personal friends, made a Craniological Museum surpassing in extent the united collections of half of Europe, and one which must now be consulted by every scholar before he can undertake to write upon the great questions involved in the natural history of man.[42]

Morton was notably one of the first people who Louis Agassiz journeyed to meet upon his arrival in the United States. In Morton, we are told, Agassiz found no less a scientist than he had in Cuvier.[43]

There is a convergence of the work that took man as the measure of all things and the work on the measurement of this relatively new creature. With Morton we find a movement that takes us from the examination of the surface of the body – as in phrenology, in the classification of skin colour and hair types, and in the determination of the beautiful form – to the interior which deepens from the space of the crania to mapping the gene. The scientific ideology of race is not about the phrenological surface – although exterior appearance and measurement remain central to classification – but about the interior where we should find the natural essence that will settle 'the grandest question' once and for all. 'With the publication of *Crania Americana*, however, the impossibility [of differentiating races with scientific precision] became accomplished fact'.[44] Morton attempted to evade the entire question of equality by first avoiding explicit commentary on the species question, and then arguing that some 'families' such as the Negro exhibit an internal diversity that corresponds to the entire range exhibited by humans, or at least possess 'a singular diversity of intellectual character'.[45] A few Negroes and mulattoes might be just as smart as a European, Morton speculated, which it must be acknowledged was more than some would admit. As Jefferson had noted, some mulattoes might contribute to the building of the nation. Now the question was, which ones? To Morton's followers, the place to answer the grandest question was America.

> There are reasons why Ethnology should be eminently a science for American culture. Here, three of the five races, into which Blumenbach divided mankind, are brought together to determine the problem of their destiny as they best may, while Chinese immigration to California and the proposed importation of Coolie laborers threaten to bring us into equally intimate contact with a fourth. It is manifest that our relation to and management of these people must depend, in a great measure, upon their intrinsic race-character.[46]

At the time Morton's *Crania Americana* was published, phrenology was the leading scientific theory for the prediction and diagnosis of pathologies and degeneration. The scope of George Combe's massive *Phrenological Remarks* captured the extent to which phrenology had become an accepted subject for scientific debate and experimentation. Still, even before Morton, the opposition to phrenology was noteworthy, even if it was overwhelmed by the popularity of the new science. Emerson remarked on the phrenological craze in his essay 'Experience'

> I know the mental proclivity of the physicians. I hear the chuckle of the phrenologists. Theoretic kidnappers and slave-drivers, they esteem each man the victim of another, who winds him round his finger by knowing the law of his being, and by such cheap signboards as the color of his beard, or the slope of his occiput, reads the inventory of his fortunes and character. The grossest ignorance does not disgust like this impudent knowingness ... I saw a gracious gentleman who adapts his conversation to the form of the head of the man he talks with! I had fancied that the value of life lay in its inscrutable possibilities ... I see not, if one be once caught in this trap of so-called sciences, any escape for the man from links of the chain of physical necessity. Given such an embryo, such a history must follow. On this platform, one lives in a sty of sensualism, and would soon come to suicide.[47]

On the other side of the Atlantic, Hegel devoted a portion of the *Phenomenology of Spirit* to the critique of physiognomy and phrenology: 'Individuality has now become the object for observation, or the object to which observation now turns'. This is the result of the failure or inability of psychological observation to understand individuality as a unity, i.e., as 'what its world is, the world that is its own ... [a] unity whose sides do not fall apart ... into a world that in itself is already given, and an individuality existing on its own account'. After an extensive rebuke of physiognomy on this point, Hegel turns to phrenology and its assertion that 'the actuality and existence of man is his skull bone'. Hegel rejected even a moderate form of phrenology that allowed the processes of the brain to have some effect on the 'skull-bone'. Some might take this position on the mistaken grounds that 'when a conscious mode of Spirit had its feeling in a specific area of the skull, the shape of this part of the skull might perhaps indicate what that mode is, and what is its special nature'. There are many who claim to feel a 'painful tension' in their heads when deep in thought, when committing murder or when writing. Phrenology would hold that the 'area of the brain' from which this feeling erupts is 'more moved and activated' and alters the nearby portion of the skull. It does not matter whether this process is passive or the result of active 'sympathy or consensus'. In either event the skull would have to prove itself 'not to be inert', and proponents must show that activity alone could stimulate 'the adjacent area of the skull-bone' to 'enlarge or diminish itself or modify its shape' in 'sympathy or consensus' with the kind of mental process or feelings that stimulated its transformation.[48]

Phrenology fails, Hegel tells us, precisely because the skull cannot feel, and that any 'feeling in the head' could simply be the result of many other factors. A feeling is itself indeterminate and it is almost impossible to distinguish one feeling from any other. There is in phrenology, Hegel argues, a fundamental reduction of the 'multitude of mental properties' to the phrenologist's deterministic and narrow understanding of mental properties. In phrenology, the multitude of mental properties 'become fewer and ... more detached, rigid, and ossified, and therefore more comparable with them'. So it is that a 'multitude of other properties' are ignored by phrenologists just as they might ignore the insignificant bumps on a murderer's skull. 'The skull of the murderer has – not this organ or even sign – but this bump. But this murderer has as well a multitude of other properties just as he has other bumps, and along with the bumps also hollows; one has the choice of bumps or hollows.' Hegel's contempt for phrenology drips from the page. Phrenology is at bottom illogical and irrational:

> The observations indulged in sound as sensible as those of the dealer and of the housewife about rain at the annual fair and on wash day. Dealer and housewife might as well make the observation that it always rains when a particular neighbor goes by, or when they eat roast pork. Just as rain is indifferent to circumstances like these, so too from the standpoint of observation, a *particular* determinateness of Spirit is

indifferent to a *particular* formation of the skull. For of the two objects of this observation, one is a dry, sapless *being-for-itself*, an ossified property of Spirit, the other is an
equally sapless *being-for-itself*; such an ossified thing as both are is completely indifferent to everything else. It is as much a matter of indifference to the high bump whether
a murderer is in its vicinity, as it is to the murderer whether flatness is close by him.[49]

At this point in the text, Hegel did grant the possibility that the phrenologists could indeed be on to something significant. A phrenological theory that
severely restricts itself to limited statements about particular facts is possible, i.e.,
the connection of a particular bump with the particular passion of a particular
individual observed over time. But,

> natural or everyday phrenology – for there must be such a 'science' as well as natural
> physiognomy – already goes beyond this restriction. It not only declares that a cheat-
> ing fellow has a bump as big as your fist behind his ear, but also asserts that, not the
> unfaithful wife herself, but the other conjugal party, has a bump on his forehead.
> Similarly, one can imagine the man who is living under the same roof as the murderer,
> or even his neighbour, or going further afield, imagine his fellow-citizens, etc. with
> high bumps on some part or other of the skull, just as well as one can *imagine* the
> flying cow, that first was caressed by the crab, that was riding the donkey, etc. etc ...
>   When, therefore, a man is told 'You (your inner being) are this kind of person
> because your skull-bone is constituted in such and such a way,' this means nothing else
> than, 'I regard a bone as *your reality*'. To reply to such a judgement with a box on the
> ear, as in the case of a similar judgement in physiognomy mentioned above, at first
> takes away from the *soft* parts their importance and position, and proves only that
> these are no true *in-itself*, are not the reality of Spirit; the retort here would, strictly
> speaking, have to go the length of beating in the skull of any one awaking such a
> judgement, in order to demonstrate in a manner just as palpable as his wisdom, that
> for a man, a bone is nothing in *itself*, much less *his* true reality.[50]

Samuel Morton, while agreeing with many critiques of phrenology that Hegel
gave voice to, was not so much interested in debunking it as in putting its search
for difference on firm scientific ground by confirming its diagnoses through
an examination of the crania itself. In truth, despite the support of respected
phrenologists such as Combe for Morton's research and their agreement on the
importance of the crania, their differences can be found in the movement from
studying the surface of the specimen to the measurement of its interior. Morton
did not look at the surface of the skull, but turned the skull over and filled it up.
Capacity was more important that shape, at least at this level of analysis. 'To the
American statesman and the philanthropist, as well as to the naturalist, the study
thus becomes one of exceeding interest'.[51]

The removal of the skin and flesh allowed the skeleton to be examined as
the organism's 'habit', a term long used in botany and natural history for the
characteristic growth pattern of the limbs of a plant or tree. Especially in winter
when there is a lack of identifying leaves and flowers, habit marks the plant as

a specific, particular, species. In terms of the human animal, its marks are not superficial expressions but describe its underlying structure. Morton argued that if one wants to understand this structure, one must go deeper than the superficial expressions of the skull and begin to measure the brain as well. This was Morton's attempt to recover phrenology, but it was an attempt that ultimately turned him in a completely different direction. In one sense, this move by Morton could be understood as his tearing away of the superficial to expose the strata of the body – skin, flesh, muscles, tendons, nerves, etc. – down to the bedrock of the skeleton.[52] We find this same movement from the surface to the interior in the development of objects and instruments from the caliper to the intelligence test that are used to distinguish between groups or races.[53]

Crania Americana

Morton based his investigation of crania on as many as thirteen separate measurements of each skull. He was interested most in cranial capacity as a measure of the size of the brain. Morton's methodology, exact and detailed, is scientific, though flawed. In fact Morton's notes were so detailed that Gould could replicate his studies and repudiate most of his results. Gould notes that Morton's errors were not deliberate attempts to deceive, and the existence of his laboratory notes supports this conclusion.[54] Morton simply did not see that he was altering his results. Like Jarvis, Morton had unconsciously put forth error but, unlike Jarvis, he did not recognize his errors.

Morton accepted Blumenbach's classification of human races, which he took to be more scientific than Camper aesthetic judgements, although it is true that Blumenbach remained convinced that beauty was a limited measure and that Caucasians set the norm for beauty. Morton's work does represent a break with the aesthetics of human variety in favour of a true scientific ideology. He abandoned the notion that human variety is the result of adaptation to climate when the antiquity of human races made this an impossible conclusion.

> One might still insist that climate could change color, but who could say that climate could change the osteological character of the individual, determine the capacity of the skull or the angle the face make the horizontal? Here was a portion of the human frame which, protected from the rays of the sun and the fetid air of the marshes, could not be influenced by environment. Here was the line of demarcation between the old anthropology and the new: mathematical measurement was supplanting aesthetic judgment.[55]

Morton wanted to avoid religious controversy and so had to be prodded to argue explicitly against the authority of the biblical account of creation. Instead he argued that the biblical chronology made the case for the antiquity of the races all the stronger since time and climate could not account for the diversity of humans. That this same evidence could be read as having undermined and contradicted the biblical story of common origin was not stressed by Morton but it was an inevitable conclusion he willingly left to his readers. Stanton describes him at the beginnings of his career at the same time that Darwin was beginning his voyage. Where Darwin would be relatively isolated from scientific circles for five years, Morton found himself in the centre of scientific discourse.

> In 1831 Morton was elected corresponding secretary of the Academy of Natural Sciences, a position which, through the scientific reputation that the Academy was acquiring in these years, brought him into correspondence with the leading scientists of many nations. His custom of holding a 'weekly soiree' to which he invited friends and 'strangers distinguished in the various departments of learning and philosophy,' increased his already wide acquaintance. A lucrative practice, together with an inheritance received upon the death of his uncle in Ireland, enabled him to live with some ease and to finance publication of his books. Tall, cadaverous, and 'of a large frame, though somewhat stooping,' with 'bluish, grey eyes, light hair, and a very fair complexion,' of an urbane, though somewhat retiring nature, Morton was an altogether improbable person to foment revolution in American science, to provide the boots and saddles and spurs with which to ride the mass of mankind.[56]

George Squire observed that, although Morton was 'essentially a man of no theories, he brought to the service of science an earnest love of truth in its simplest and severest form … He had, in short, a true appreciation of the dignity and aims of philosophy.'[57] As his initial studies on crania were being organized, Morton began a substantial correspondence with George Gliddon, then the American consul in Egypt. Gliddon had already developed a strong interest in Egyptology and had an equally strong interest in the question of the unity or plurality of human species. In 1830, Morton and Gliddon began their mutually beneficial friendship. Gliddon's appreciation for Morton was clear from the beginning, and Morton's affection for his 'young, restless, and argumentative protege' was clear.[58] Their letters discuss the proper arrangement classifications and the degree to which this would turn on the proper identification of human variety. While serving as the American Consul in Egypt, Gliddon bought and stole a number of human skulls and shipped them to Morton in order to further his research. Charles Pickering remarked how Gliddon had obtained a mummy case he just happened upon while travelling through Egypt, preventing it from being used as kindling by a local resident.[59] It was Gliddon it should be remembered who introduced John C. Calhoun to Morton's work and in 1844 Morton dedicated his book *Crania Egyptia* to Gliddon.[60]

Gliddon distinguished himself as a popularizer of Egyptology whose public lectures drew hundreds of listeners each night. In Boston and New Orleans, Gliddon's lectures included a stage set with an 800-foot-long moving backdrop of panoramas and illustrations which depicted historical events, geographic regions and specimen as he discussed them. For seven nights in Boston, he lectured on various aspects of Egyptian civilization, and at the end of each lecture, removed a layer of wrapping from a mummy said to be that of a female Egyptian aristocrat. It did not diminish his rising fame that on the final night the mummy upon examination turned out to be rather obviously male.[61] Gliddon dropped this portion from later lectures on the tour in favour of artefacts and mummies whose anatomy was already known.

Gliddon's lectures remained widely popular for years and resulted in his being presented as a character in Edgar Allan Poe's short story 'Some Words with a Mummy' as one of a trio of learned gentlemen attending a private unrolling of an Egyptian mummy. Gliddon's character translates the hieroglyphic inscription as the name of the mummy: 'Allamistakeo'. Upon unwrapping Allamistakeo and finding him still well preserved, it is decided to apply electric shocks to the mummy in hopes of it responding to the stimulus. It does respond, at first with a seemingly involuntary shaking of its fist in Gliddon's face. Much to the amazement of the assembled guests, Allamistakeo regains consciousness and immediately admonishes Gliddon for his views on Egypt. Gliddon makes himself invisible, literally so Poe tells it, but later reappears only to be told by Allamistakeo, whose real name is Count, that he has mistakenly interpreted Egyptian archaeological remains just as he has its language. Ultimately, the Count's point is that his society has dealt already with the many problems of civilization that we believe are new to our age. He is a historian and was embalmed alive so that he might be awakened whenever the historical record which he and others like him zealously guard has become corrupted.

> An historian, for example, having attained the age of five hundred, would write a book with great labor and then get himself carefully embalmed; leaving instructions to his executors *pro tem* that they should cause him to be revivified after the lapse of a certain period – say five or six hundred years. Resuming existence at the expiration of this time, he would invariably find his great work converted into a species of hap-hazard note-book – that is to say, into a kind of literary arena for the conflicting guesses, riddles, and personal squabbles of whole herds of exasperated commentators. These guesses, etc., which passed under the name of annotations, or emendations, were found so completely to have enveloped, distorted, and overwhelmed the text, that the author had to go about with a lantern to discover his own book. When discovered, it was never worth the trouble of the search. After re-writing it throughout, it was regarded as the bounden duty of the historian to set himself to work immediately in correcting, from his own private knowledge and experience, the traditions of the day concerning the epoch at which he had originally lived.[62]

The Count believes in fixity no less than his modern interrogators. His periodic existence is dedicated to preventing 'our history from degenerating into absolute fable', such as that of the most recent theories of human origins. He is shocked to learn the interlocutors believe that the creation occurred 'only about ten centuries before' he was entombed. No one during his time would have 'entertain[ed] so singular a fancy as that the universe (or this world if you will have it) ever has a beginning at all'. The polygenic theory is also familiar to him, he says, but the mistake is to read the Egyptians as literally meaning five different creations and types of man. It was meant figuratively, as denoting social rank, and not as a theory of 'the origins of the human races'.[63] The Count recalls 'a man of many speculations' who 'remotely hinted' at the subject. He too referred to Adam, but the word was not a name but the phrase 'Red Earth'.

> He employed it, however, in a generical sense, with reference to the spontaneous germination from the rank soil (just as a thousand of the lower genera of creatures are germinated), – the spontaneous generation, I say, of five vast hordes of men, simultaneously upspringing in five distinct and nearly equal divisions of the globe.

No one in his day would have seriously considered such a thing. At this, those present shrugged their shoulders or 'touched their foreheads with a very significant air' at the suggestion that the polygenic theory is itself a degenerate 'absolute fable' of the Adamic account it had been used to tear down. Swift Buckingham,[64] another guest at the unrolling, glances 'slightly at the occiput and the sinciput of Allamistakeo' and speaks of phrenology and mesmerism as advancing the theory that we must 'attribute the marked inferiority of the old Egyptians in the particulars of science, when compared with the moderns, and more specifically with the Yankees, altogether to the superior solidity of the Egyptian skull'. The Count responds that in his day, too, 'the prototypes of Gall and Spurzhem had flourished and faded ... and the maneuvers of Mesmer were really very contemptible tricks ... Great movements were awfully common in his day, and as for Progress, it was at one time quite a nuisance, but it never really progressed.'[65] It was no accident that the polygenic theory of his time supported the social order, just as it was most perfect to the social order of antebellum America. Through a range of fields, the Count points to equal achievements in his own time. The mummy is only bested and made to admit that progress had been made when he is confronted by our two most profound advances: laxatives and patent medicines for the treatment of venereal disease.[66] It is nice to note that Poe's interest in Gliddon and the Egyptology fad was not merely abstract. The spread of Egyptian Revival architecture during this same period was a style associated with the tombs of the wealthy and with the spaces of government. The Tombs prison in New York City and the entrance to the Moyamensing prison in Philadelphia are great examples of the latter association, while the former can be found on a walk through any cemetery dating from the time.[67]

Gliddon was not so timid in life as he was in fiction. Privately, Nott regarded Gliddon as a rascal and not at all a serious scientist. But together he and Gliddon went one step further than their mentor in debunking religious authority. In doing so they were working out tendencies in Morton's work, as they demonstrated by including Morton's unfinished manuscript on the 'Origin of Human Species' in their *Types of Mankind*. It was presented as Morton's final statement on the species question, and is one that explicitly stepped into some of the controversies that Morton had avoided much of his career. The manuscript is fragmentary and ends abruptly, but in many ways it is an expanded version of his published 1842 lecture 'Hybridity in Animals, considered in Reference to the Question of the Unity of the Human Species'. To Nott and Gliddon, the manuscript provided sufficient evidence of the continuity across Morton's works with those of his students and acolytes writing after him.

Morton's lecture drew on scientific and anecdotal accounts of hybridity in wild and domesticated plants and animals. Morton noted that species are not necessarily distinguished by their lack of interfertility. The relationship of hybridity to species formation was far more complex than the simple ability to produce offspring – the viability of the offspring was what was really in question. If the racially pure infant or child fell victim to a mortal illness, the cause was not to be looked for in the parentage, but if a 'hybrid' infant or child died of the same circumstance, the cause was to be found in the mixture of the types.[68] Viability would come to have a specific meaning beyond simply surviving one's birth.

And yet, Morton noted, while it was easy to find examples of hybridity, it was nevertheless assumed to be a great rarity or even a monstrosity. The same nature that made hybridity possible also made hybrids a rare event. 'While we admit that hybrids, as a general law, are contrary to nature, we are also compelled to concede that this law has very many exceptions'. What is it that prevents hybrids from being more common? They are common enough for Morton to imagine that left unchecked, the 'animal world should present a scene of confusion'.[69] He approvingly notes Prichard's argument that 'there is some principle in nature which prevents the intermixture of species, and maintains the order and variety of the animal creation'. The natural principle which hybrids violate by their very existence is nothing less than 'the natural repugnance between individuals of different kinds'. As monsters had in previous centuries, the hybrid became evidence that the order of nature had been violated. Although these violations are found in nature, they are more often found in and encouraged by civilization as a 'state of domestication, in which the natural propensities cease, in a great measure, to direct their actions'.[70] In transforming our own 'natural propensities', domestication at first opposes nature but ultimately frees us from nature's laws. These natural laws were replaced by the judicial administration of populations, social hygiene, and disciplines. However, these areas of social life were of no great concern to Morton except for the debilitating effects of domestication. Morton argued that the need to administer the population of the nation had arisen because slavery had lowered the barrier of 'natural repugnance' of the African for the European, and vice versa. The social relation of master turns the European away from proper moral behaviour, especially in sexual matters. Sex and slavery corrupt the sexual mores of the master, Morton noted, and as the slave can be assumed to lack morality, the natural repugnance of each species for the other is too easily overcome by the desire for pleasure. In such circumstances, there will be a proliferation of hybrids seemingly producing even more variety.

To understand fully 'the moral degradation consequent to the state of slavery', Morton urged us to recognize the significance of domestication as an evolutionary, i.e., developmental, force:

> that domestication [in plants] evolves the faculty of hybridity there can be no question, and we would apply the principle to various classes of animals. We have shown that this fact is unquestionable among some quadrupeds and some birds, of which the hybrid varieties have been cultivated for the uses of man.

Morton introduced a new 'law of nature' from the ruins of natural repugnance: the law of domesticity. The capacity for crossing varies across species 'in proportion to their aptitude for domesticity'.[71] Not only does domesticity destroy the natural order, it actively encourages hybrids and increases their fertility. Taboo, fertility and viability were no match for the greater forces of the law of domesticity and cultivation. If we work hard at maintaining it, a social repugnance may be substituted for the natural one, but Morton thought that in the end slavery would dissolve the natural repugnance between species into a mere moral prohibition that vice all too often overcomes. Civilization conquers the species barrier only to imperil itself.

And so Morton could find hybridity everywhere and still only see monstrosities and violations of nature. The urge to cross – or urge for sexual satisfaction – is an urge that is usually channelled by desire and natural repugnance, but these limits were overcome by the intervention of humans. While not every species is capable of change, humans change because we domesticate ourselves, Morton suggested. Since hybridity is everywhere regulated by domesticity, Morton argued that interfertility could not serve as a test for speciation in humans. Because domestication 'evolves' the capacity for hybridization, human hybrids are not surprising nor do they violate the general law that 'hybrids are counter to nature'. The theory of a 'natural repugnance' returns again as a means to understand social phenomena: 'the repugnance of some human races to mix with others has only been partially overcome by centuries of proximity, and, above all other means, by the moral degradation consequent to the state of slavery'.[72] Morton believed that Africans and Europeans equally abhor each other, but slavery drew the two together and provided the opportunity for crossing. Morton argued that slavery was an ancient bond providing us with evidence of its importance for understanding the diversity of human origins, but that now slavery was leading to a confusion of types.

Underneath or behind, but sometimes obvious to the reader of Morton's writings on hybridity and the species question, is his fundamental belief that degeneracy is fostered by civilization and that domesticity already threatens the society that it produces. In the order of nature, no other animal has 'the aptitude for domesticity' except man, who is 'as Blumenbach expressed it, the most domestic of animals'.[73] There is nothing surprising about the pronounced 'power of fertile hybridity' given our aptitude for domesticity – the plasticity demonstrated by hybridity was interpreted by Morton as a morbid result of domesticity. The fertility of human hybrids, far from proving the unity of mankind, now proved the power of domesticity to overcome nature's highest barriers. The mulatto showed the dangers of domestication and the threat of hybridity, but it also endowed the slaver with the respectability of the bourgeois domination of nature and social life.

The existence of hybrids became evidence for the proposition that 'the human family should prove to embrace several distinct species', now that 'fertile reproduction has ceased to be evidence of identity of species'.[74] The proof was to be found in Egypt, which Morton explored through his study of his collection of Egyptian skulls, *Crania Egyptiaca*. Morton relied on his empirical studies of his crania collection and on the authority of others. He confirmed his observations of the crania using some of the cultural artefacts of Egyptian civilization. Like Cuvier, whom he frequently cited, Morton believed that the representations of racial types in Egyptian art confirmed his cranial evidence of the antiquity of the races and gave additional credence to the possibility of separate origins.

Morton followed the publication of his *Crania Egyptiaca* with a short notice of a second series of crania that had come into his possession through the efforts of William Gliddon, George Gliddon's nephew, 'of Cairo, who, prompted by that extraordinary interest in Egyptian questions which seems inherent in his family, has availed himself of every opportunity for extending our knowledge of the people and the monuments of the valley of the Nile'.[75] This new series confirmed the findings of the first, although there were difficulties which Morton carefully acknowledged. The Prussian Scientific Commission had recently opened several well-plundered tombs that were difficult to date due to the mixing and jumbling of the contents during repeated ransackings. For Morton, the abundance of these 'desecrations in the search for treasure' inadvertently revealed the antiquity of the remains. The presence of Greek inscriptions, though, does 'not make them Greeks'. More was needed, and it was the crania that could verify who left these remains and artefacts. The crania had first to be classified using what Morton termed 'an ethnographic analysis of this series'.[76] Putting aside the arguments over the antiquity of the find, Morton noted that what was not in dispute was that a skull of Indo-European linage could be easily identified no matter where it might be found.

Morton's ethnographic analysis suited his the available evidence: the shape of the head, the size and placement of the teeth, the capacity of the crania and the facial angle. Two other important factors, Morton sadly noted, were unavailable to him, namely characteristics of the hair and 'the bony meatus of the ear'. Morton asked rhetorically if

> Nilotic physiognomy ... indicated [by] monumental and sepulchral evidences ... [of] the small, long, and narrow head, with a somewhat receding forehead, narrow and rather projecting face, and delicacy of the whole osteological structure ... [a] mean adult [facial] angle ... greater than that given by the large series measured in *Crania Egyptiaca*? Is this owing to the fact, that the heads now under consideration belong to persons of distinction and probably therefore, of education and refinement?[77]

Morton does not provide a direct answer, but sometimes the asking of a question is as revealing as any answer, as the question here indicates nothing other than the possibility that a connection exists between the shape of the head and one's education and refinement. Likewise the one intact 'Pelasgic' skull in his sample is simply 'perfect, and well characterized in most of its proportions'. No more description was deemed necessary than this. The most numerous type in his new collection, and one that was also to him the most uniform in appearance, was the Semitic head which is described only as having more 'the common Arab than the Hebrew cast of features'. The Negroid skull is problematic due in part to its variation. As Morton had already noted, his new series contained two Egyptian/Negroid hybrids, and his single Negroid specimen was 'a mixture of the Caucasian form and the Negro form, in which the latter predominates'. This Negroid had the facial angle of the Pelagic head, but the smallest brain capacity and largest jaws of all the others. Despite such problems, which Morton took to be aberrations in the head itself, he concluded that these new crania 'sustain in a most gratifying manner' the results of his first study of 100 skulls, and he gladly amended his 'ethnographic table ... to embrace all the ancient Egyptian skulls now in my possession'.[78]

In the manuscript which Nott and Gliddon published posthumously in *Types of Mankind*, Morton gave perhaps his most sustained discussion of 'the problem of the origin of the human species'. From his relatively early *Crania Americana* until this posthumous essay, Morton usually shied away from contentious disputes and controversy. While he was fully aware of the consequences and logical conclusions of his cranial studies, Morton appeared at times reluctant to admit them. When he did, it was often with a hesitancy his acolytes never quite understood and certainly never achieved. Even in the exchanges with his monogenist nemesis Rev. John Bachman, he often appeared more pained by the need to defend his scientific analysis than adopting a belligerent tone. Morton did not take an active interest in either supporting slavery or in the undermining of biblical authority. It was with 'diffidence and caution' that he had touched on the growing controversy around his work and that of his acolytes. It is perhaps only in this posthumous work that he fully states his views on the polygenic theory. It is in a 'very few remarks' on this 'zoological question and one inseparably associated with classification in Ethnology' that he finally summarizes twenty years of 'education and reflection'.[79]

All the while that he protested his allegiance to the biblical chronology, Morton constantly pushed the creation further and further back into the past in order to protect the Mosaic account while showing its limits. He presented the problem with the biblical chronology as squarely set within a larger context of 'the original plurality of races' and avoided the more limited question of the plurality of origins. With this subtle shift of emphasis, Morton avoided what his followers would embrace, but it was not lost to anyone that they all shared areas of fundamental agreement regarding the polygenic theory of human origins. If we do share a common origin, Morton offered, then we must come up with a better record of this event, especially as we must admit that not enough time has passed to populate the world with its many human varieties. Certainly, Morton argued, it must at the very least be admitted that we need a new chronology of the history of the earth that coincides with our knowledge of the history of man. This new chronology would be founded upon the temporal and spatial distribution of the human species. The choice was made clear by the challenge of the polygenists: accept monogenesis and you have to reject the possibility that a rational deity or intelligent designer brought the world into existence; but accepting polygenesis means that you replace the creation of every man 'in God's image' with an order in which the Caucasian is the original, most intelligent, beautiful and God-like of the types of mankind.

Natural history had tried hard to maintain the link between the species question and classification, which, could now only be maintained if one assumed a plurality of origins. The 'commonly received opinion', i.e., the opinion derived from biblical authority – that all humans have a collective origin in a 'primeval pair', rests upon a broader assumption that the variation in human appearance, type of civilization, language and a host of physical and moral differences is the result of the sudden appearance of 'accidental varieties whose survival through isolation and intermarriage have rendered their particular traits permanent'. Morton followed the authority of Prichard in viewing the 'manifest inadequacy of the influence of climate, locality, civilization, and other physical and moral agents acting through long periods of time' to explain the obvious 'diversity of nature'. That monogenic writers had more or less modified and blended together these forces was the result of tradition rather than scientific and sceptical investigation.

> Guided in this inquiry solely by the evidence derived from nature ... we would be led
> to infer that our species had its origins not in one, but in many different creations;
> that these were widely distributed into those localities upon the earth's surface as were
> best adapted to their peculiar wants and physical constitutions, and that, in the lapse
> of time, these races, diverging form their primitive centres, met and amalgamated,
> and have thus given rise to those intermediate links in organization which now con-
> nect the extremes together.[80]

The theory that could rationally connect the wide range of data would be nothing other than the 'direct study of man himself' and the comparison of man
with other members of the zoological series. Finally Morton and his followers
would establish the meaning of variety. And they did, until Darwin reinscribed
the notions of accidental variety and survival that Morton had dismissed.

If Darwin's theory was already 'in the air' then certainly Morton breathed
that same atmosphere although they started from very different places. As will
be mentioned in the next chapter, Darwin begins with the assumption that genealogy is the basis for classification of life. Morton followed Cuvier in assuming
that geographic range is evidence of special creations. Cuvier's revolutions were
of a general planetary – almost Hesiodic – scale, at least in his early discussions
of revolutions.[81] Morton's creations occurred on a more modest scale. Humans
were 'distributed' into localities he termed 'primitive centers', each with unique
or specialized plants and animals. The distribution of human types along with
complementary flora and fauna could only suggest special creation and design.
The five primary races or as Morton notes 'more appropriately ... groups' are
themselves evidence of five different creations in distinct 'primitive centers'.
Whether these creations were simultaneous or representative of the order of
divine nature became the question at hand once the polygenic origins of humans
had become the established theory.[82]

Armed with this new interpretation, Morton and his followers argued that
any continuity between a 'smaller or greater number of primary races' is the
result of amalgamation. The primary races have 'an indigenous relation to the
country they inhabit', and a 'collective identity of physical traits, mental and
moral endowments, language, etc'. If we had descended from a single ancestral
pair, then one would assume that the human species would appear more or less
the same across the earth and even the further back in time we go in search of
the 'single aboriginal type'. However, the farther back we go, Morton argued, the
more we must admit the great antiquity of the races. The doctrine of the fixity of
species could not be fully reconciled with the idea of adaptation; adaptation was
only necessary at the moment of creation when the creator adopted the plant or
animal to its geographic region. Fossils still occupied a limited place in polygenic
science, mostly as evidence for revolutions of the earth. If species are fixed, then
fossils cannot tell us anything about contemporary species. What could it matter

if one assumed that there is no change over time and no genealogical connection? To Morton, the diversity found at the end of history varies little from the original creations. The races appeared separately. To the naturalists of the time, evidence from Egypt, China, Eturia and elsewhere established long-standing and fundamental differences between the five races.

> It is not necessary to multiply authorities upon this point; for it is worthy of remark that every philologist of distinction who has investigated the subject, has arrived at precisely the same conclusions; although few have ventured to make public the deductions to which they inevitably lead. The doctrine of a diversity of origin in the human race, although gathering supporters daily, has yet so few open advocates, and is generally esteemed so radical a heresy, that investigators in this, as in many other departments of science, hesitate in pushing their researches to their ultimate results. The discussion of the question cannot, however, be long postponed, and it is not difficult to foresee in what manner it will be finally determined.[83]

To Morton and many other naturalists philology proved the polygenic theory despite the reluctance of anthropologists to admit 'so radical a heresy'. The question of the diversity of human origins could not be postponed given its inherent link to that other question: the origin and perpetuation of slavery and the fixity of the master/slave relation. Morton challenged the Jeffersonian belief that amalgamation might lead to the elimination of the lesser traits of the lesser parent because the morbid law of hybridity made such improvement difficult. After all, Morton noted, those that questioned the law of hybridity were mostly 'French and Spanish writers ... and it must be remembered that the Spaniards and a certain proportion of the population of France are themselves already as dark as any quinteroon, or even quadroon'. Amalgamation might more easily happen in France and Spain, where 'any few crosses would merge the dark into the lighter race' than those places populated by a purer European stock. It was a different matter in the United States, where 'the amalgamation [of the European and the Negro] is going on upon an immense scale'. Here Morton believed, 'mulattos, or mixed breeds', would 'die off before the dark stain can be washed out by amalgamation ... it is a hard matter to wash out blood'.[84]

The misleadingly named *Types of Mankind* was the most widely read work on polygenism, and with the help of *DeBow's Review* and other publications, helped to popularize the theory. The title is misleading because the book is really a discussion of the species of mankind with contributions by Agassiz, among others. *Types of Mankind* laid out the general polygenic propositions of the American School to an appreciative and receptive scientific and general readership (see Figures 2.2 and 2.3). And there was little doubt on the part of the authors regarding the originality of polygenism. It claimed to be no less than a revolution in our understanding of ourselves.

It will be observed that, with the exception of Morton's, we seldom quote works on the Natural History of Man, and simply for the reason that their arguments are all based, more or less, on fabled analogies, which are at last proved by the monuments of Egypt and Assyria to be worthless. The whole method of treating the subject is herein changed. To our point of view, most that has been written on human natural history becomes obsolete; and therefore we have not burdened our pages with citations from authors, even the most erudite and respected, whose views we consider the present work to have, in the main, superseded.[85]

It is not difficult to trace the correlation between the theory of multiple biological origins and the theory of multiple 'homelands' for language, as well as the theory of multiple birthplaces of civilization. The origins of races coincided with the origins of language, with a distinctive organization of social space that is emphasized by the continuity of language. 'The two authors have aimed to construct a theory of human natural history from purely scientific and archeological evidence' said Haven in his review of American anthropological science.[86] The polygenic theory became the accepted theory of the origins of human variety.

> The question of the human race, whether a *unity* or not, is being now discussed, with great ability, by naturalists all over the world. We may mention among others, Morton, Pritchard, Nott of Mobile, whose contributions have appeared in our Review, and who has lately written an able work upon the subject, Bachman of Charleston, also the author of a late treatise, and Professor Agassiz. This subject has an important bearing just now, in examining the position occupied by the Negro, whom philanthropy is seeking to elevate to the highest *status* of humanity. It is something to discover that he is not of common origin with the Caucasian, and this appears to be the better opinion among scientific men.[87]

Like Morton, Louis Agassiz had an ambivalent relationship with biblical authority. He would consistently throughout his entire career in America affirm this authority and the fixity of species by special creation while arguing for the polygenic origins of humans. He was and still is acknowledged as the most serious scientific adversary of Darwinian theory. Agassiz first made his reputation in Europe with his analysis of fish, basing his work in Cuvier's theories. After some failed business dealings in tandem with an extravagant lifestyle led to the inevitable sufferings of a young gentleman living beyond his means, Agassiz emigrated to the United States. Such a move took him from the reach of his creditors and allowed him to pursue the 'free scientific inquiry' that America seemed to offer him. Soon after this relocation, Agassiz was confronted with the importance of race for the solution of the species question. With all the authority of his professorship at Harvard, Agassiz argued with great determination for fixity, design and polygenesis. Agassiz's interest in the species question was stimulated by two events. The first was his adaptation of Cuvier's theory of successive revolutions as an explanation for fossils. The second was his encounter with Morton. Accord-

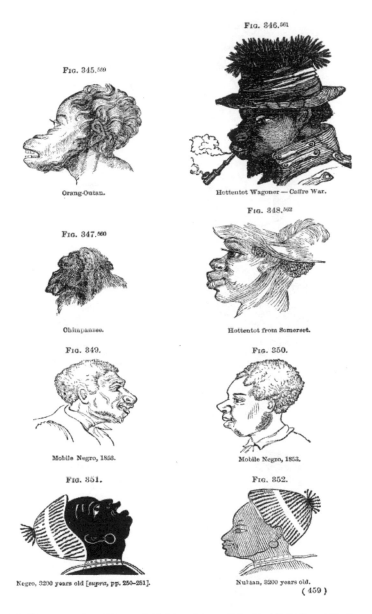

FIG. 345.⁵⁵⁹

Orang-Outan.

FIG. 346.⁵⁶¹

Hottentot Wagoner — Caffre War.

FIG. 347.⁵⁶⁰

Chimpanzee.

FIG. 348.⁵⁶²

Hottentot from Somerset.

FIG. 349.

Mobile Negro, 1853.

FIG. 350.

Mobile Negro, 1853.

FIG. 351.

Negro, 3200 years old [*supra*, pp. 250–251].

FIG. 352.

Nubian, 3200 years old.

(459)

Figure 2.2: Illustration from Nott and Gliddon, *Types of Mankind* (1855), p. 459. From the author's collection, 8th edn (1860).

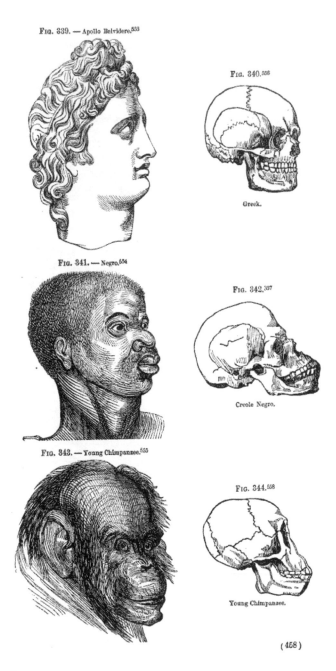

FIG. 339. — Apollo Belvidere.[553]

FIG. 340.[556]

Greek.

FIG. 341. — Negro.[554]

FIG. 342.[357]

Creole Negro.

FIG. 343. — Young Chimpanzee.[555]

FIG. 344.[556]

Young Chimpanzee.

(458)

Figure 2.3: Illustration from Nott and Gliddon, *Types of Mankind* (1855), p. 458. From the author's collection, 8th edn (1860).

ing to Lurie, Agassiz had little interest in race, nor was he deeply troubled by the implicit support for slavery that lay with the polygenic theory. His first interest was in free scientific inquiry, and this was not to him a disingenuous or ironic statement. The American School believed strongly in the need to free science from the dogma of religion. For this to become a reality, the biblical chronology had to be superseded by a scientific chronology that would give a deeper understanding of Christianity. However, it is difficult to believe that Agassiz was deaf to the nearing drums of war or to the fiery abolitionist and rabid pro-slavery orations in Boston, Philadelphia and Charleston.[88]

Agassiz was notably frank in describing his experiences in his newly adopted country. Before coming to America, Agassiz had rarely considered and racial classifications not relevant to much of his own early work. He wrote that he had never seen a Negro until he arrived in the United States. Once acquainted, he immediately became convinced of the inferiority of Negroes because of what he described as his 'profound visceral revulsion' at being served by them in his Philadelphia hotel. Gould did one of his many great services to the history of science by printing Agassiz letter in its entirety for the first time:

> It was in Philadelphia that I first found myself in prolonged contact with Negroes; all the domestics in my hotel were men of color. I can scarcely express to you the painful impression that I received, especially since the feeling that they inspired in me is contrary to all our ideas about the confraternity of the human type [*genre*] and the unique origin of our species. But truth before all. Nevertheless, I experienced pity at the sight of this degraded and degenerate race, and their lot inspired compassion in me in thinking that they are really men. Nonetheless, it is impossible for me to reprocess the feeling that they are not of the same blood as us. In seeing their black faces with their thick lips and grimacing teeth, the wool on their head, their bent knees, their elongated hands, their large curved nails, and especially the livid color of the palm of their hands, I could not take my eyes off their face in order to tell them to stay far away. And when they advanced that hideous hand towards my plate in order to serve me, I wished I were able to depart in order to eat a piece of bread elsewhere, rather than dine with such service. What unhappiness for the white race – to have tied their existence so closely with that of Negroes in certain countries! God preserve us from such contact![89]

It was during this time that Agassiz first met Samuel Morton, whom Agassiz recognized immediately as a scholar who was 'after Georges Cuvier ... the only zoologist who had any influence on his mind and scientific opinions'.[90] It is not surprising then that Agassiz contributed a long essay to the great summation of the American School that took distribution and fixity as its central theoretical assumptions. The surface of the earth could be read like the crania of the degenerate; the geographic bumps and spacings made sense of the races and their differences. This is Agassiz's debt to Cuvier through extinction and change (revolution). Agassiz's essay anticipates his long-running dispute with Darwin and

Darwinism. Agassiz's reply to those who believed, like Buffon and Pliny, that nature is in constant flux was simple. Variation, according to Agassiz, must not be thought of as occurring over time, but across space. In an attempt to preserve natural history, Agassiz had in fact fatally undermined it because he removed history from consideration. Left only with nature, science had to either posit a series of world ages or eras, or allow materialism its day and admit chance and time as factors in the formation of the earth and its inhabitants.

Some commentators have taken Lurie's view that the polygenic theory 'simply rested on the assumption that the Creator had ordained the existence of separately and specifically created human species'.[91] While this might be true most famously of Agassiz, the coalition of naturalists, physicians, philologists and archaeologists that comprised the American School was much more interested in that goal of modern science: free inquiry. The polygenic theory demonstrated the necessity for such free inquiry and the need to separate science from religion. 'The truth will out', as Agassiz himself had written. The opposing monogenic view was defended by Bachman, Lawrence and Wells 'in a manner that was in some respects markedly anticipatory of Darwinian evolution'.[92] However, there is usually a silence on what specifically in these theories anticipated Darwin. This fact should serve to remind us that Darwin was not working in isolation, but was engaged in the most important scientific dispute of his time. Lurie perhaps inadvertently reminds us of how history is a most fickle set of narratives. The name of John Bachman, for example, does not appear in his list of monogenists, yet the Charlestonian naturalist and clergyman played a pivotal role in the monogenic/polygenic debate as Morton's most distinguished critic before Darwin. The implication is that the polygenic view was less than scientific. This is perhaps because the polygenic view 'anticipated' Darwinian evolution. The notion of a creator's guiding hand on the great chain of being was held by no less a polygenist than Agassiz, but he came late to the debate when the polygenic theory had already been framed. His contribution to the *Types of Mankind* appeared well after polygenic theory had dispensed with serious consideration of the biblical chronology. Moreover, monogenic views – no matter how much one would like to read them as mere precursors to Darwin – rested on biblical authority as well as on a different section of Genesis. The contradictions between the two creation stories in Genesis were obvious, but these contradictions supplied both sides with the means to claim divine and scientific authority. Both Morton and Bachman could agree that at some point human variety constituted its own limit.

John Bachman, the great adversary of polygenism before Darwin, is perhaps best remembered now for his collaboration and friendship with James Audubon. In the Jeffersonian tradition, he too devoted his efforts to education, science and service. Bachman worked tirelessly on a wide range of activities, all the while keeping up a busy ministry. He founded the small liberal arts Newberry College in Newberry, South Carolina. Clearly natural history held a special importance for Bachman. Teaching at the College of Charleston, he joined the 'Circle of Naturalists'. Charleston became one of the centres of natural history, attracting such notables as Agassiz and Audubon. Bachman's association with Audubon began through an almost chance encounter and grew into a lifelong friendship, including the marriage of two of Bachman's daughters to Audubon's sons. They spent a great deal of time together collecting along the coastal plain of South Carolina and disputing their divergent ways of life. Bachman often admonished Audubon for his love of 'gog and wine and snuff'. Maria Martin Bachman, who had married Bachman after the death of her sister – Bachman's first wife – found in Audubon's friendship the opportunity to become a notable natural history painter and illustrator. Her contributions included the beautifully and accurately detailed entomological drawings that accompany Audubon's birds. This she learned by studying the abundant specimens available to her as well as the drawings in *Entomology of North America*. One of her descendants noted in *Charleston Receipts* that 'It was she who painted many of the backgrounds for Audubon's famous paintings'.[93] All this while she gave birth to fourteen children and suffered from *tic douloureux*, an extremely painful chronic nerve disease. Audubon wrote of her that 'Miss Martin with her superior talents, assists us greatly in the way of drawing; the insects she has drawn are, perhaps, the best I've seen'.[94] Unfortunately, many of her paintings and sketchbooks were lost in the destruction of Charleston and its aftermath.

Audubon readily acknowledged his debt to John Bachman as well.[95] Bachman supplied most of the text for *Viviparous Quadrupeds of North America* and it was Bachman and Audubon's sons who published the book, Audubon having already died. Bachman was no mere amateur, as Morton and Agassiz would at times claim, but one of the more recognized figures in the history of science in America. That Bachman is now remembered best as an amateur naturalist, a clergyman or as the founder of Newberry College is one legacy of the triumph of the polygenic theory. Yet Bachman's name remains preserved as a footnote to book dealers' descriptions of Audubon's work, and in the names Audubon gave to several species: *Haematopys bachmani* (black oystercatcher), *Sylviagus bachmani* (brush rabbit), *Sciurus niger bachmani* (eastern fox squirrel), *Vermivora bachmani* (Bachman's warbler, now extinct), *Aimophila aestivalis bachmani* (Bachman's sparrow), and the *Libythaea bachmani* (American stout butterfly).

While the scientific acceptance of the polygenic theory often coincided with the political acceptance of slavery, the convergence of the two was not always certain, It would be easy to over-generalize the relation between slavery and the monogenic/polygenic debate. Supporters of monogenic theories were inclined to disapprove of slavery for theoretical reasons, but were no way uniformly convinced of the equality of the races. Common origin did not necessarily translate into equality and might actually accommodate difference.[96] Slavery was a key referent in the debate between the monogenists and the polygenists, but the debate over the origins of human diversity would have happened anyway and indeed had been well under way in its modern form since the nineteenth century. 'It was obvious to anyone of that period that human evolution would be the battle ground on which the question of transmutation would be fought out'.[97] Bachman was a formidable adversary of Morton and in the course of his defence of the monogenic origin of humans at times approaches Darwin's arguments regarding adaptation to changing circumstances. But he was prevented from realizing Darwin's insight by his refusal to allow chance in nature and by his defence of slavery.

Perhaps more than anyone, Bachman captured in his work the contradictions that animated the monogenic/polygenic discourse: as a confirmed supporter of the Union, he opposed succession, but he also supported 'the institutions of South Carolina' despite his defence of monogenism. In 1860, John Bachman was listed as owning four slaves and his wife one.[98] At the same time, Bachman trained the first three black Lutheran ministers, one of whom would later become president of Wilberforce University and another the Chief Justice of the Liberian Supreme Court. Free Blacks constituted 35 per cent of Bachman's congregation in 1860, and the segregated Sunday school for blacks was larger than the one for whites, with 150 students and a staff of 30. Services were accompanied by a three-piece band, leading some to suggest that it was one of the sites for the origin of American church music. But it was nonetheless a segregated church, enlightened only in relative terms. As Bachman had declared at the beginning of the war, he knew 'the duties and dangers of the moment'. The church's newly installed bell was melted down to provide shot for the confederate defenders as Sherman's army of liberation approached. Sherman's troops destroyed the building housing the Sunday school and library, and along with it Bachman's papers and collections. The native of Rhinebeck, New York, did not welcome the forces of the Union, but rather fled inland. Captured by Union scouts, he was beaten and left with a paralysed arm.[99]

Bachman did not simply give the opening prayer for the secessionist South Carolina legislature and then retire from politics, as some of his apologists claim, in a sort of loyalty to South Carolina and silent opposition to the division of the Union. Bachman marshalled the results of his scientific labours to the cause of his beloved state. In his essay 'The Duty of the Planter to his Family, to Society, and his Country', Bachman implored the planters and manufacturers to heed his call to devote their production to the materials necessary 'to feed and clothe and cheer the heart of the soldier, now in the camp or on the battlefield'. Bachman wrote that his recommendations were from one 'scientifically and practically acquainted with the cultivation' of these products. 'The dangers and duties of the planter' required the recognition that 'every product necessary to the support and comfort of man ... can be easily produced on your own plantations and farms'. The conclusion of the piece leaves little doubt about Bachman's political sympathies:

> Husbands and sons are in the army – their wives and children must be supported. If you join the extortioners in high prices, you will effect our subjugation by your avarice, and bring destruction on our beloved Southernland. We are all engaged in the same noble, patriotic, and Christian cause.[100]

Bachman signed his brief survey of what he now termed 'military Natural History' as Curtius. Marcus Curtius was a Roman noble who, upon hearing that an enormous chasm had opened in the Forum that the seers said could only be closed if the most precious Roman possession was tossed into it, realized that this referred to the bravery of its citizens, and rode his horse fully armoured into the abyss, thus sealing the rift. An altar was left to mark Curtius, and Bachman was himself buried under the altar of his Lutheran church.

If there is one morbid science, it is polygenesis and its descendants, degencracy and criminal anthropology. If such a view seems harsh, or if the origin of race still seems so distant to us today, it is only because we are still shaped by the pride and prejudice that prevents our disciplines from acknowledging their descent from these earlier fields. And there was great nationalist pride in the origins of these sciences.

> It is not without an emotion of national pride that we hold up to public view the works whose titles we have above recited. They are all American, all relate to a subject of the highest importance, all are works of original investigation, and of high scientific character, and together form a valuable contribution to the stock of human knowledge. They discuss the grandest problem of natural history, the question of the unity of the human race.[101]

The continuity between the problems of natural history and the problems of the sciences of life and society are not always so easy to trace, and it is an exaggeration to claim that the polygenic theories led inevitably to the rationality of fascism, but it is not an exaggeration to note that they were the result of a scientific ideology obsessed with race, crania and degenerates. The search for crania was not merely a pre-Darwinian activity, nor were the practical opportunities for cranial studies that furthered the classification of human variety confined to the practice of tomb-robbing and popular lecturing. The criminal anthropologists, the sociologists who espoused eugenic reformism, and the scientists and physicians who participated in the *Ahbenaber* did not simply draw inspiration from the American School: they desired the same knowledge. This continuity can be found in a notice printed in the *Journal of the Anthropological Institute of New York* in 1872:

A communication to the German Society of Anthropology during the past winter invokes the attention of all persons interested in science to the importance of making use of the opportunities for ethnological research furnished by the war between France and Germany; and the author, while acknowledging the difficulty of attending to such matters during military operations, expresses his earnest hope that every possible effort may be made to secure a good series of skulls and brains of the African tribes brought by France into the conflict, and especially those of the Turcos. We have not yet heard to what extent this suggestion was heeded by those who had the opportunity.[102]

In the twentieth century, we can still find a continuity that is difficult to ignore even if it is easy to exaggerate. In an essay from the July 1909 issue of the *Annals of the American Academy of Political Science and Social Science*, Alexander Johnson wrote:

McKim in his book *Heredity and Social Progress* declares we must eliminate the degenerate by a humane and painless death – have some pleasant lethal chamber into which they may be introduced, lie down to happy dreams and never awaken. It is not worth while discussing that, not even as an academic discussion, it is so tremendously far away.[103]

3 DARWIN IN CONTEXT:
SCIENCE AGAINST SLAVERY

The continuities with which we ended the last chapter show, if there is any question, that the Civil War and emancipation did not end the monogenesis/ polygenesis debate. Instead, its end came with Darwin's deployment of natural selection as a theory of both continuity (genealogy or common descent) and discontinuity (modification, adaptation, extinction, selection) – a deployment that shattered the natural history of man upon which polygenesis rested. Linné's scheme of classification, because it was founded on the assumption of fixity, makes sense only when allied with Cuvier's natural history i.e., the fixity of species within a period and place was allied with the marked differences between places on the Earth as well as the differences between the ages of the earth.

As with the ibis and the Negro, the key referent was the place of humans and human varieties within the natural order represented by the various systems of classification. 'Nature is man writ large, and man nature writ small' was a common ideology of the day and one could not be faulted in venturing to say that it is an ideology that we continue to share. While a rigid classification could not account for change, more imaginative classifications did allow for the possibility of variation within species. What remained was the exploration of the mechanisms that would allow for change over time and yet remain consistent with the needs of a rational system of classification. Of course, Darwin does this through selection (both natural and sexual) and genealogy. With the discovery of natural selection, natural history reached a crisis that furthered its transformation into biology. This crisis was not over the possibility of disciplinary change, but over the place of humans in the natural order. This remains today much as it was in the past a political as well as a scientific question whose stakes were made clear by Darwin's decisive intervention into the monogenic/polygenic debate.

Over time it has become uncritically accepted that the ideas and concepts Darwin brought together so masterfully in *The Origin of Species* had been 'in the air' as part of the 'spirit of the age'. But was everything already neatly in place and pointing to the same inevitable conclusion? We do know that Combe's *Phrenological Travels in America* sold 350,000 copies while *The Origin of Species* sold about 50,000 in Darwin's lifetime. Was Darwin's work the mere assembling and making intelligible insights already available? So we are often told and so too did those contemporaries like Smith accuse Darwin of not only failing to acknowledge his own sources, but their own individual discovery of the process of speciation.[1] It is a commonplace to read how 'Darwin slid into place the final piece of an enormous puzzle where other wonderful minds had played. Natural selection had been in the air, waiting to be born.'[2] What might be termed the 'atmospheric thesis' assumes that a general scientific consensus had been reached and that Darwin supplied the capstone. Instead, we find that this is precisely the moment when natural history had reached a crisis amidst the disputes over fixity, variation and classification. If a puzzle was before Darwin, it had been laid out by the polygenists because their theories had become the accepted norms of scientific ideology. If the history of science involves the history of forgetting, then here, in the transition from natural history to the study of life that is thought to be so seamless, we can witness such an error of forgetting.

Darwin purposely avoided the use of the terms evolve or evolution until the very last sentence of his book in order to avoid any confusion of his work with the already well-known use of the term.[3] Evolution at the time of *The Origin of Species* was most often used in the sense of an inevitable and determined unfolding over time of characteristics already present from the beginning. The homunculus was not a curiosity or error, but one of the best examples of this type of evolutionary view: 'fantastic as it may seem today, all future generations had been created in the ovaries of Eve or testes of Adam, enclosed like Russian dolls, one within the next – a homunculus in each of Eve's ova, a tinier homunculus in each ovum of the homunculus, and so on'.[4] Darwin redefined evolution to mean change over time, i.e., change directed only by the needs of the individual to survive its struggle for existence and the ability of the species to adapt and vary in the course of the struggle for life. Instead of a movement towards an end or a higher stage, with Darwin the history of nature became the struggle of life to perpetuate itself in part through the preservation of 'slight changes'.

It is without a doubt true that theoretical arguments regarding the meaning of evolution were already a part of scientific discourse, as well as present in the works of history, political economy and morality. Wallace's near simultaneous 'discovery' of natural selection is used to bolster the oft-repeated view that evolution and natural selection were 'in the air'. The use of individual biography as a determining factor in the history of science is a somewhat shallow form of analysis. It is certainly true that the relationship or lack thereof between Darwin and Wallace is superb evidence for understanding scientific discovery as a collective and disciplined production of knowledge. It is also true that if not Darwin then Wallace, if not Wallace, then someone else – if the social conditions mandated the search for origins, which they of course did, then the search for origins would be central to scientific inquiry. A concern for the scientific understanding of human origins became the sign that designated an anthropological study as scientific. Certainly, the specific articulation of the theory derived from Darwin's account of his experiences would differ from others. If nothing else, it would be a great loss not to have Darwin's excellent prose. The greater loss would have been that Wallace ultimately did not want to apply natural selection to the study of human evolution, whereas for Darwin this was essential to his refutation of the polygenic theory. Likewise, the bestowing on Mendel of the title Father of Genetics after five decades of obscurity, when his little-read work on heredity was just another candidate for the gnawing criticism of the mice until quite literally rescued from a wastepaper basket.[5] A scientific biography is overdetermined by the scientific ideologies of the subject's era. Moreover, it does not necessarily follow that just because Darwin and Wallace came to the same conclusion that they arrived at this conclusion by the same path. The production of knowledge at any time creates an intricate web of rational scientific truths and seemingly rational scientific ideologies.

Changes in any discourse can have profound implications for others, but the continuities appear alongside equally significant discontinuities. Taking, for example, the various precursors and contributions to Darwin's theories – if one asks the question 'What were the different fields where we can see the groundwork for Darwinism?' there are three that were noted by Canguilhem. First, the fossil, taken as the marker of the origin also became the measure of time and the record of change and discontinuity. Second, embryology had come to provide substantial evidence for a common course and common origin to all living things. Life really had begun to be a universal vital connection which genealogy would make indissoluble. As a result, recapitulation and preformism dominated evolutionary speculation, but as Darwin would hint in the structure of *The Origin of Species*, recapitulation became for him a summary of accumulations and of earlier conditions of life which serve as the basis for future adaptations and not a determinist cycle. Third, rudimentary organs presented evidence of both descent from earlier forms and of change over time. From this, it is but a small step to conclude as well that the variation found in nature results from descent from a common ancestor and not the degeneration of a primitive type. From this moment forward, classifications could be only provisional arrangements of the forms of life. Instead of geographically independent creations, nature could be presented as a genealogical tree encompassing all living things (see Figure 3.1).[6]

Man has been studied more carefully than any other animal, and yet there is the greatest possible diversity amongst capable judges whether he should be classed as a single species or race, or as two (Virey), as three (Jacquinot), as four (Kant), five (Blumenbach), six (Buffon), seven (Hunter), eight (Agassiz), eleven (Pickering), fifteen (Bory St. Vincent), sixteen (Desmoulins), twenty-two (Morton), sixty (Crawford), or as sixty-three, according to Burke. This diversity of judgment does not prove that the races ought not to be ranked as species, but it shows that they graduate into each other, and that it is hardly possible to discover clear distinctive characters between them.[7]

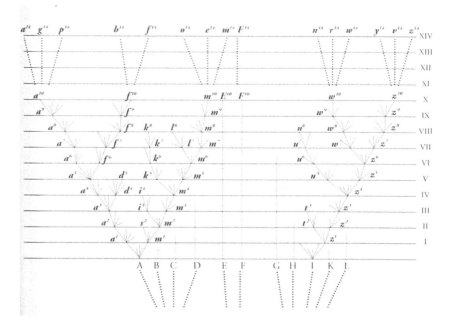

Figure 3.1: Darwin's Genealogical Sketch from the *Origin of Species*. Reproduced with permission from *The Complete Work of Charles Darwin Online*, ed. John van Wyhe (http://darwin-online.org.uk/).

Before descent with modification could be accepted as a scientific norm, it had to be distinguished from the general formation of scientific ideologies in which it was embedded. In this sense, descent with modification resolved the question of the fixity of species by eliminating fixity as a scientific problem. Fixity was not a problem of an abstract theoretical discourse, but a production with which one might finally define man. On the one hand, Darwin put to rest the scientific discourse on the problem of origins as dominated by the monogenic/polygenic debate. Darwin's work was grounded not only in the elements that he carried forward – the importance of the fossil record, comparative morphology, embryology and rudimentary organs – but also in the debates and discourses which he would either transform or destroy. On the other hand, Darwin had to distinguish his monogenist theory from those common amongst Christian theologians and abolitionists.

We have already encountered much of this pre-Darwinian formation in four general tendencies which become over time much more noticeable. First, there was a general movement from the exterior to the interior of the object of knowledge. This can be found in the movement from phrenology to craniology, in the movement from criminal anthropology to genetic determinism, or in the short distance separating the 'Bad Seed' and the defective gene. Conversely, in the case of medicine, one can also find the movement away from the body through the mediation of technology and multiplications of positions that come between doctor and patient and in the profusion of medical specialties.

Second, the authority of the biblical chronology had already been successfully refuted or at the very least called into doubt by the polygenist American School who accepted the scientific reality of an ancient and changing Earth. Cuvier had maintained that the biblical flood was the historical record of the latest revolution of the earth, and that the Earth has had many more spread over vast periods of time. The permanence and antiquity of the races appeared to Morton, Nott and others as no less confirmed by archaeology than was the antiquity of the Earth confirmed by geology. Lyell had showed that the processes that formed the Earth are still at work today, and though he hoped almost until the end that these processes would reveal the hand of a designer, gradual geologic change convinced him that the Earth was so ancient as to be almost beyond comprehension. Cuvier accounted for the complexity of the fossil record by introducing the concepts of extinction and revolution, though he did so only to bring into alignment the biblical chronology with the mounting evidence against it. From these elements it would be easy to conclude that the Earth changes, either through revolutions or gradual change, and with it species either adapt or are driven to extinction. Fundamental questions were constantly in play: if a species is lost, what happens to its 'place in nature'? Is a new species created or must one admit to there being gaps or discontinuities in the fabric of nature that life is always in a struggle to fill? Extinction explained the fossils and, using this same explanation, the fixity of species could be maintained, but extinction also worked against fixity, for the Earth was neither filled by one species, nor – except at its beginning – ever empty of life, despite the countless past extinctions.

Third, the application of human variation as a model for understanding all other populations of animals and plants settled a fundamental question: could the order and variations of human species shed light on the natural order? The variation of humans and variation in nature as a whole could not be the result of special creation or a series of creations, but of gradual transformations, the preservation of small differences and mere chance mutations (the term others still used was 'sports'). These processes were explained as being the result of intelligent design. Darwin's intention was to do just the opposite when he proposed that variation was inherent in nature itself. There could be no opposition between law and variation when variation is a law not of the 'planets spinning by', but of the life teeming on this one.[8] Freed from design, nature now stood as an end in itself which valorized it, but also allowed for a kind of separation between the natural and the human that would further the course of the domination of both.

Much of the monogenic/polygenic debate concerned the discourses of hybridity, and so the debates explicitly referred to slavery and the hierarchy of 'the races' along with the need for free inquiry. These debates contributed to Darwin's work because they were as much within the accepted norms of scientificity as other allied fields such as embryology and comparative anatomy which constituted the fourth contribution to Darwin's work. One would, or course, not expect to find it strange that slavery was also 'in the air' at the time, and indeed scientists sworn to explain the origins and meaning of human life and variety conducted their work with reference to slavery. It would perhaps be more surprising and far stranger if they made no reference to it on the eve of the American Civil War.

It was not lost on anyone at the time that they were witnessing a transformation of scientific knowledge – that the spirit of the age was the new structure of knowledge and its disciplines. The praises of progress masked the insecurities and the dislocations of knowledge that constitute the destructive aspects of 'progress'. Physics and chemistry were already becoming the province of specialists; the laboratory was becoming the locale for organizing the production of scientific knowledge. The rapid foundation of new learned associations and societies reflected both the move towards specialization and the speedier dissemination of results and theories. Science had finally begun to study life as it is lived, i.e., life at the speed of life.[9] Just one governing principle remained to be overthrown: the view of man as the apex of creation. For Agassiz and others, man retained his special place in nature. Even if he was not to be thought of as having been placed everywhere from the beginning, he was still destined to rule creation. *The Origin of Species* is a profound argument for human humility. Before Darwin, the history of the Earth could be identical with the history of man, but it was now possible to assert that humans were key to understanding the history of life in relation to the history of the Earth.

Here the work of Darwin's mentor, Charles Lyell, deserves attention. Lyell's *Principles of Geology* argued that the evidence of past changes in the surface of the Earth can be explained by processes still in operation. Though a friend of Darwin – and the broker of the publishing agreement between Darwin and Wallace – Lyell had great difficulty accepting the conclusions of *The Origin of Species*. Nonetheless he ultimately did after taking up the species question in a series of notebooks written between November 1855 and November 1861. His own *Journals on the Species Question* documents his struggle with this fundamental question.[10]

Lyell's difficulty was not so much with natural selection, but with the conclusion that nature produces life without any need for a 'creative intelligence' or 'a supreme artificer'. In his *Journals*, Lyell attempts a last stand against his young friend's emerging theory of natural selection and chance as the real 'powers of nature'. The advocates of natural selection simply 'deify matter and force'. If true, then there should be a space for one who imagines a creative intelligence having a human will, a 'personal deity designing the evolution of new forms ... a supreme being, the variety-making power ... delegated by a supreme artificer'.[11] In some ways, this is Lyell's attempt at the deification of natural selection. But Darwin did not allow himself to indulge in such self-deception. He has the same manner of the keen-eyed observer that he admired so much in Lyell, who might sit watching the ocean waves erode away a shoreline or a stand of goldenrod as numerous insects arrive, interact and depart with food if of course their stopping has not provided a meal to others. Darwin's own garden at Down House contained in miniature the tangled bank he saw across all of the natural world. Lyell knew that his position was a weak response to Darwin, but he was determined to seek the 'creative intelligence' he stubbornly insisted moves through the doctrine of descent. Of course, he could only do this by ignoring or significantly minimizing the role that Darwin allowed for chance.

Lyell did not discount the significance of descent. He found momentary comfort in the possibility that it might finally prove the great chain of being: 'the tie by descent may prove to be the hidden bond which connects general orders and classes together of the Animal Kingdom & eventually as a true theory which no one can reasonably doubt'. Natural selection in and of itself does not rule out the possibility of 'creation in successive periods'. 'It is probable that we do not so much overestimate the human as underrate the organic world'. This does not exclude the possibility of 'creation in successive periods of new beings'.[12] It is at this point that Darwin injects chance into the scheme of nature. Darwin allowed for the possibility that development might be rationally directed, but saw that chance intervenes to upset any plan. In addition to this, sexual selection and social knowledge also intervene in the creation of new varieties.

Lyell used the apt phrase 'permanent variety-making machinery'[13] to preserve the mechanistic analogy while attempting to reconcile it with the relentless working of chance. A creator was no longer needed to provide either similitude or difference, nature itself was organized by design to produce variety permanently. Like any believer, Lyell could still hold on to the hope that a creator's hand rested on the controls of the machine of life. His metaphor is a mystification of chance in nature, like his *Principles of Geology* took naturalists in the direction of material interpretation. The problem lies in a 'machine' which has a permanence and fixity that is at odds with the concept of 'permanent variety-making'. The machine produces variety without changing itself. Though Lyell eventually accepted the explanatory value of natural selection, he steadfastly refused to abandon fixity. And so his machine of nature restated the metaphor of fixity.[14] As a metaphor the machine is transposed into nature itself so that nature and the machine function as one.[15] This machine is no less revealing as a metaphor than it is as a direct statement of fact. Lyell presented what might be the last defence of a mechanistic nature. Darwin took Lyell at his word and worked out some of the logical implications of such 'machinery'. He recognized that the force behind the 'permanent variety-making machinery' was nothing less than chance itself. This proposition puts Darwin in the great tradition of materialism that can be traced back through to Lucretius and Epicurus. A machine of nature would be a very odd contraption in any scheme of a nature that is indefinite and infinite, and so Darwin's two metaphors of the tangled bank and of the tree of nature became for him representations which he knew his readers would understand even as he radically redefined them. In so doing, Darwin juxtaposed the genealogical descent of all living things to the spiritual bonds of a great chain of being.

To represent the affinities of all beings by a great tree of nature has an obvious biblical reference and Darwin transformed the medieval Tree of Life into a tree of genealogical affinities. 'I believe this simile largely speaks the truth' Darwin wrote with characteristic modesty.[16] The medieval Tree of Life, however, was an expression of the great chain of being. Darwin's tree is not a giver of life but a history of life: a representation of a dynamic and indeterminate nature characterized by continuity, discontinuity, descent and extinction. Where Enlightenment taxonomies depended on fixity of species, Darwin's genealogical tree 'weaves a network of kinship between all beings'[17] no matter how different they might now appear. To quote the passage from *The Origin of Species* in full:

> The green and budding twigs may represent existing species; and those produced during former years may represent the long succession of extinct species. At each period of growth all the growing twigs have tried to branch out on all sides, and to overtop and kill the surrounding twigs and branches, in the same manner as species and groups of species have at all times overmastered other species in the great battle for life. The limbs divided into great branches, and these in to lesser and lesser branches, were themselves once the former and present buds by ramifying branches may well represent the classification of all extinct and living species in groups subordinate to groups. Of the many twigs which flourished when the tree was a mere bush, only two or three, now grown into great branches, yet survive and bear the other branches; so with the species which lived during long-past geological periods, very few have left living and modified descendants. From the first growth of the tree, many a limb and branch has decayed and dropped off; and these fallen branches of various sizes may represent those whole orders, families, and genera which have now no living representatives, and which are known to us only in a fossil state. As we here and there see a thin straggling branch from a fork low down a tree, and summit, so we occasionally see an animal like the *Ornithorynchus* [the platypus] or *Ledidosiren* [South American lungfish], which in some small degree connects by its affinities two large branches of life, and which has

apparently been saved from fatal competition by having inhabited a protected station. As buds give rise by growth to fresh buds, and these, if vigorous, branch out and over-top on all sides many a feebler branch, so by generation I believe it has been with the great Tree of Life, which fills with its dead and broken branches the crust of the earth, and covers the surface with its ever-branching and beautiful ramifications.[18]

Language could not accurately depict nature, and no total taxonomy is possible when nature is seen in its infinite variety. After all, Darwin had written as early as 1842 that those who knew the language of natural history and zoology were actually a rather small group and already in constant contact with each other. Because of their specialized language – a language that Darwin, Henslow, et al. wanted to preserve for them – they had no way of communicating their ideas to a wider audience, and the 1842 report on classification already displayed little enthusiasm for the possibility of a comprehensive classification. In the report on taxonomy, one hears clearly the scepticism of Darwin, but also the optimism of his peers regarding the possibility of just such a comprehensive taxonomy. The situation might have seemed chaotic at best and it was true that science was bur-dened with 'the vagueness and uncertainty of its nomenclature', but the problem was not the 'diversities of classification' because if we give 'perfect liberty to the systematists' they will take us to the 'true system of nature' that is the ultimate goal of natural history.[19]

While it is commonly agreed that the ground for Darwin was 'already laid' by others, the nature of Darwin's intervention against the scientific ideologies of his day is not always explicitly discussed. The American School was certainly a part of the respectable scientific establishment; the concepts of fixity and design had by now returned to the centre of scientific debate. In addition, the view of evolution as the development or the unfolding of a destiny united works ranging from phi-losophy to embryology. Finally, the scientific support for already existing slavery crystallized in the production of knowledge. It was this social structure that Dar-win completely undermined. After quoting Spencer's *Social Statics*, Singer wrote that 'the *Origin* came thus to a world well prepared' by the work of Spencer. But Singer erred in claiming that Darwin mistook the concept of evolution as inter-changeable with 'the species question'.[20] The most pressing issue was not evolution *per se*, but the definition of evolution.[21] The dispute was marked by questions of whether species exist, if species change and, if they do, what their origins and eventual destinies are. If they do not change, then the varieties of humans could be classified as different races or species. If this was true, then was there really a scientific, rational enlightened argument against imposing slavery and terror on those lesser creations? Would not the Enlightenment principles of universal his-tory and progress at best necessitate the tutelage of slavery for the lesser races?[22] Too often seen as anti-religious, Darwin's scientific intervention was no less dev-astating in its implications for the domination of nature and humans.

The Origin of Species in the Context of Slavery

One must have a theory of the fixity of species in order to create a truly comprehensive system of classification. When Darwin introduced chance and infinity of variety into the debate on the species question, the possibility of creating such a comprehensive classification lost its earlier promise. In time, classification came to be replaced by process and flows of energy replaced intelligent design as natural regions became ecosystems. Darwin placed less emphasis on definition and classification in favour of drawing the attention of his peers to the importance of selection, relations between individuals of the same species in addition to relations between species. In doing so, nature could be competitive, mutualistic, sexual and genealogical as opposed to merely a war of all against all. The sciences of society and nature represented nature and man as one. Humans became the observers and the products of processes which do not depend upon them but to which they are indebted. At the same moment that life became the object of knowledge, man was displaced from a privileged place in the natural order.[23] Here we can find an obvious instance of Darwin's 'transvaluation of all biological values'.[24] Darwin presents life as engaged in a dynamic and indeterminate struggle for existence. This dynamic is expressly built on relations of struggle, mutualism and sex. As a result of our grounding nature in such social relations, 'the *social* reality of nature and *human* natural science or the *natural science of man*' became 'identical expressions'.[25] This view originated in natural history, as we find in Cuvier's revolutions of the Earth, but with Darwin's intervention these aspects of natural history took on fundamentally new meanings.[26]

The Origin of Species is structured as a single argument with digressions to address anticipated objections of critics. Though 'hastily' written, Darwin had already outlined the argument some years earlier in an as-yet unpublished manuscript. The reasons for his sudden haste to publish have been discussed by Darwin's biographers without a definitive conclusion, but we do have the text, whether written in haste or not. It begins with an exposition on variation as it is found under domestication and then in nature without the intervention of humans. Morton had also noted the importance of domestication and 'domesticity' to what he saw as the many violations of a law of hybridity. Darwin postponed the discussion of hybridity until later, thereby separating the question of the *effects* of domestication from the *frequency* and *viability* of hybrids. Instead of Morton's generalization of fixity, Darwin's discussion assumes variation as a generality that emerges from individuals and local events. Individuals of the same species vary across time and space. Their struggles are local ones, as in the plant at the edge of the desert that struggles for existence. One constant is that variation is itself the condition for and the essential product of the struggle for life. The process of natural selection becomes the basis for understanding this variability. Life, embroiled as it is in the struggle for existence, maintains itself through variation.

The remaining portions of *The Origin of Species* are given over to anticipating objections to the theory of natural selection and descent with modification. The first problem Darwin mentions concerned instincts and is especially focused upon slave-making ants and the mutualism of aphid/ant relationships. Only then is hybridity discussed as being essential to the permanent production of variety, and not as a violation of fixity. Other problems of the geological record (fossils, catastrophe and extinction), the succession of organic beings (pre-formism, teleology) and geographic distribution (design and special creation) followed as additional areas from which objections might come. Instead of merely revising accepted theories, Darwin presented a new science structured around a classification based on genealogy, morphology (the comparative study of function, behaviour and environment), embryology and the study of rudimentary organs. That he did not give Lamarck sufficient credit must have been at least partly due to a desire to get his views a fair hearing by distancing himself from Lamarck's discredited theory. Darwin might not have defended Lamarck, but he was much closer to Lamarck than he was to the polygenists. The structure of *The Origin of Species* thus reveals to the informed reader a transvaluation of natural history into the study of life: 'All true classification is genealogical, that community of descent is the hidden bond which naturalists have been unconsciously seeking, and not some unknown plan of creation, or the enunciation of general propositions, and the mere putting together and separating objects more or less alike'.[27] Darwin's genealogical tree of evolution represents the degree of continuity and discontinuity present in the history of nature. Species might appear as expressions of continuity, but they are also evidence of past extinctions and gradual adaptations. Human history is not the striving of different species for supremacy, but the conflict within one species as it confronts the conditions of life on the earth. Nature and the struggles between species and groups of species is not a state of permanent war and it does not suffer revolutions that can be compared to social revolutions. Darwin frequently noted that torments and pains are rare and brief in the natural world. Contrary to other forms of life, humans have learned to make suffering itself into a way of living; one need only observe the torments of animals under the whip or the knife of the vivisectionist, the horrors of modern warfare or the enslavement of other humans. The willingness of humans to inflict torment and pain that flows through each of these acts might be one of our unique attributes, but slavery, vivisection and war were all of great concern or actively opposed by Darwin. Anyone who engages Darwin's work knowing only the current social controversies which invoke his name is immediately taken aback by his manner of presenting his argument. It is not as it is with the reading of Nietzsche, where one must look for the ironic qualities of his polemics. Nonetheless, the intention here is not so much to explain how the misuses and misunderstandings came about, but to describe the conditions that made Darwin's work so revolutionary. There is a forgetting of the profound relevance of slavery to the species question in the appropriation as well as the denunciation of Darwin. Darwin was not at odds with many of his predecessors, or even with the religious authorities, but with the scientific ideology of his day. Despite Hyman's predisposition to read Darwin (and others) as literature with a capital L, his comments do elicit from the texts Darwin's uses of metaphor and the slippages of language that only reinforce the distance between Darwin and his two most famous 'predecessors': Malthus and Spencer.[28] Neither merit much attention in Darwin's 'Historical Sketch'. Darwin cared little for political economy – after Marx sent him a copy of *Capital* Darwin thanked him for his 'great work' and remarked that he only wished that he knew enough about 'the deep & important subject of political Economy' to understand it.[29]

Malthus supplied Darwin with much less than it might seem. It is likely that Malthus's work only supplied a necessary framework for understanding growth; it was lightly read by Darwin, whose primary method was direct observation and for whom theoretical choices were determined by the degree to which they coincide with experience and observation. Many commonplace criticisms levelled against Darwin fall flat when it is remembered that he undermined the very ideas that religion and much of philosophy deeply shared: belief in providence and purpose in nature. Nor did Darwin's use of such bourgeois theorists as Malthus and Spencer make him a supporter of a Hobbesian polity or an apostle of capital. Darwin's insistence that the struggle for existence takes place between individuals rather than between classes does put him within political economy's conventions regarding the individual, but his work is not a 'Robinsonade'.[30] Moreover, emphasizing the competition between groups would only lend support to a scientific ideology that he opposed. Darwin was not attempting to produce a political economy of nature, although the dominant ideas of his time most certainly find their way into all his writings. While he confessed that he had little interest in political economy, he was interested in classification and hesitated to emphasize conflicts between species. This was not because conflict and struggle do not exist, but because the definition of species was a more complex question than anyone believed. We had not yet settled the meaning of 'species' so it could not be taken as the location of the struggle for existence. Regardless of our ability to agree on the taxonomy, we could observe the individual plant at the edge of the desert. It is for this reason that conflict occurs between individuals and not species and could not do so until species had been defined in terms of genealogy and not political economy. The concept of the individual did not always function in *The Origin of Species* as it did in political economy. For instance, Darwin makes a lengthy note of the observation that some species produce more offspring than could possibly survive. One could, as he notes, derive the principle that the number of individuals tends to exceed the conditions of life from the actual reproductive strategy of many diverse species as we might from Malthus.[31] 'Until the one sentence of Malthus no one clearly perceived the great check amongst men.' Population increases geometrically, Darwin wrote, but crop yields fluctuate from year to year, usually with several years of bounty followed by a smaller yield and famine. However that might be with humans, 'in nature production does not increase' and so there is no overproduction to spur population growth, but populations changed in relation to populations of other species. Taking hawks as his example, Darwin wrote that on average every species must lose individuals to hawks or bad weather, and that a decrease in the number of hawks would 'affect instantaneously all the rest'. The effects of the loss of the hawks would create a 'force like a hundred thousand wedges trying to force every kind of adapted structure into the gaps of in the economy of nature, or rather forming gaps by thrusting out weaker ones'. the difference is that when the force of population works on human affairs, it not by natural selection, but 'by means however of volition'.[32]

Darwin's 'Autobiography' supplies his view of his encounter with Malthus. It was after his return from the voyage of the *Beagle* that in 1837 he began his first notebook 'on the variation of animals and plants under domestication and nature' as a means to understand variation in general. He worked on 'true Baconian principles, and without any theory collected facts on a wholesale scale'. Through the use of questionnaires, extensive reading and communications with gardeners and 'skilful breeders', Darwin came to the conclusion that 'selection was the keystone of man's success in making useful races of animals and plants'. But how did selection work in nature? What was the mechanism for selection when in nature there was no guiding hand of a gardener or rancher? '[H]ow selection could be applied to organisms living in a state of nature remained for some time a mystery' to him. Darwin marked October 1838 as his first encounter with Malthus when 'for amusement' he happened to read 'Malthus on Population'. He says that he was well prepared 'to appreciate the struggle for existence' that he found in Malthus because his own observations of 'the habits of plants and animals' showed that this struggle is consistent across all of nature.

> it at once struck me that under these circumstances favourable variations would tend to be preserved, and unfavourable ones to be destroyed. The result of this would be the formation of new species. Here then I had at last got a theory by which to work; but I was so anxious to avoid prejudice, that I determined not for some time to write even the briefest sketch of it.

He would wait until 1842 and 1844 to finally write his sketch. This encounter with Malthus had allowed Darwin to conceptualize the workings of natural selection, but Malthus did not supply enough to answer the problem that now confronted Darwin: 'the tendency in organic beings descended from the same stock to diverge in character as they become modified'. The very systems of classification, with their elaborate structures, showed this divergence without recognizing it.

> I can remember the very spot in the road, whilst in my carriage, when to my joy the solution occurred to me; and this was long after I had come to Down. The solution, as I believe, is that the modified offspring of all dominant and increasing forms tend to become adapted to many and highly diversified places in the economy of nature.[33]

Malthus supplied Darwin with a commonly known authority for a difficult proposal, and one who had already found acceptance in respectable circles, but it would be easy to overstate the influence of Malthus on Darwin.

Darwin opened up an infinite variety of possibilities by making the process of adaptation depend on *chance* rather than authority. Many of these possibilities are emancipatory, but just as many reveal themselves to be oppressive. However, the greatest number appear to be both emancipatory and oppressive. One possibility that was created by Darwin's work was the formation of Social Darwinism, but to associate Darwin with his ideological namesakes is problematic when others, e.g., Spencer, certainly loomed larger than Darwin before the publication of *The Origin of Species*, and some like Haeckel set themselves as Darwin's truest and most pure interpreters.

De Beer noted that Darwin turned Malthus's essentially anti-Enlightenment and anti-transformative limit into a mechanism for change. Malthus used his theory to argue against the possibility of change, with humans destined either to be doomed to misery and death for lack of necessities, or to be 'equally [miserable] from the effects of immorality, idleness, and sloth ... all attempts to preserve life were contrary to correct application of principle, charity was an economic sin, altruism "unscientific," and presumably the medical profession pursued an anti-social aim'. Since the possibilities of variation, shown by cultivated plants and domestic animals, were in Malthus's view strict limits, progress was impossible; attempts to achieve it, such as the French Revolution, 'were doomed to failure and mankind could neither improve nor be perfected'.[34] Darwin worked with the same evidence but in the opposite direction. He used the 'slogan-like antithesis between geometrical and arithmetical rates of increase' as an *explanation* for change and not a *limit* to it. Malthus did not 'run away from the horrors of tooth and claw [or attempt to] veil it, minimize it, or moralize on the greater resulting happiness for the survivors'.[35] The law is as important for its demoralization or even transvaluation of the struggle for existence as it is for its scientific truth. Darwin takes the same law as his opponent and uses it to achieve the completely different outcome: the support for variation and change. 'The ordinary belief that the amount of possible variation is a strictly limited quantity is likewise a simple assumption.'[36] Nothing is produced simply to harm another species, nor to help it, but each species produces together the conditions of life. This is not a moral struggle, and neither does it produce cruelty. Darwin noted that in nature the end comes quickly and without the torments that the civilized unleash on each other and on the world around them.

It was obvious that populations left unchecked increase, but this assertion had no theoretical basis, nor did it have a necessary result – struggle and deprivation – even at the level of scientific ideology. It was Malthus who supplied a theory of populations. The importance of it as an argument against fixity is less obvious, but without it as a catalyst for change, variation and extinction, 'the earth would soon be covered by the progeny of a single pair'.[37] However, fixity and monogenesis were incompatible in Darwin's theory.

> It must not be supposed that the divergence of each race from the other races, and of all from a common stock, can be traced back to any one pair of progenitors. On the contrary, at every stage in the process of modification, all the individuals which were in any way better fitted for their conditions of life, though in different degrees, would have survived in greater numbers than the less well-fitted. The process would have been like that followed by man, when he does not intentionally select particular individuals, but breeds from all the superior individuals, and neglects the inferior. He thus slowly but surely modifies his stock, and unconsciously forms a new strain. So with respect to modifications acquired independently of selection, and due to variations arising from the nature of the organism and the action of the surrounding conditions, or from changed habits of life, no single pair will have been modified much more than the other pairs inhabiting the same country, for all will have been continually blended through free intercrossing.[38]

All things being equal, populations tend to increase geometrically in relation to resources and by this Malthus meant simply ever-increasing numbers, and not ever-increasing variety. In Darwin's view any increase in population comes with an increase in variation. Rather than tending towards homogeneity, the struggle for existence serves to make the smallest variation all the more important in producing variation. Now add to this the element of chance. One could imagine coming upon the scene of a regular contest for mates in which the largest and more aggressive moose wins and chases all of its rivals away to the margins of the field, the most attentive or fit females move towards him, leaving behind their less attentive or slower rivals. Now lightning strikes the dominant male, killing him and all of the females standing closest to him. So by chance the biggest and fittest animals do not reproduce, but their less fit rivals who had been chased to the margins of the open field or not chosen by the fittest male do survive to leave offspring. Years later a previously unknown epidemic breaks out and as it happens only some of the offspring of these same less fit moose are immune, but none of the animals in other herds that descended from the brothers and sisters of our once proudly dominant moose can withstand the epidemic. Now, how is a naturalist to fix for any period of time which individual or variety is the most fit? Fitness is relative to the conditions of life, it is not an essential quality, i.e., fitness is determined by the interaction of organism with the conditions of life.[39]

In a letter, Darwin flatters Spencer by saying that until the 1870s he had never read a defence of Toryism that made any sense, but then says in another letter of the same period that he never really read much of Spencer's work and suggested to Hooker that he preferred reading the *Lady in White*.[40] So what then does Darwin get from Malthus and Spencer? Certainly, a connection to some of the most popular and influential writings of the time. But more than simply calls to contemporary authorities, Darwin gets a ready-made and accepted scientific/ economic principle. In a sense, he adapted to the intellectual environment of the time. No matter how revolutionary a work might be, it cannot exist completely outside of some intellectual discipline without risking never being heard at all. Thus, even revolutionary works display as well a certain conformity or recognition of convention and authority.

Lyell's *Journals on the Species Question* include the text of a letter from Spencer to Darwin of 22 February 1860, which Lyell's editor tells us was not in the Darwin correspondence at the Cambridge University Library at the time of publication and the original has still not been found. So we owe it to Lyell for including this notable exchange in his notebooks. Lyell characterized Darwin's notion of Natural Selection as 'H[erbert] S[pencer]'s own doctrine of "evolution" put on so satisfactory a basis'.[41] What was it that was so satisfying to Lyell about Darwin's reading of Spencer? Was it that Darwin had appropriated Spencer – though Darwin steadfastly maintained that the struggle for existence was always used as a metaphor – or that with Darwin evolution became not the *cosmic* doctrine that it was for Spencer and his disciple John Fiske, but a materialist one?[42] The acknowledgement by both Lyell and Spencer that Darwin provided just such a materialist explanation must call into question the degree to which Darwin and Spencer really coincided, as Darwin's injection of selection of spontaneous variations and rejection of design shows considerable distance between their views. Spencer's cosmic evolution could never accommodate such a doctrine of indeterminacy and spontaneity. Indeed, it has been noted that Spencer's enthusiasm for Darwin's work quickly waned, and that Darwin had little time for Spencer's views and but little more for his person. Spencer wrote that Darwin had

> wrought a considerable modification in the views I held. While having the same general conception of the relation of species, genera, orders, etc., as gradually arising by differentiation and divergence like the branches of a tress & and while regarding these cumulative modification as wholly due to the influence of surrounding circumstances. I was under the erroneous impression that the sole cause was adaptation to changing conditions brought about by habit, using the phrase condition of existence

in the widest sense as including climate, food, & contact with other organisms.[43] But you have convinced me that throughout a great proportion of cases, direct adaptation does not explain the facts, but that they are explained only by adaptation through natural selection.

Many must have been struck with the fact that among all races of organisms the tendency was for the best individual only to survive & that of the goodness of the race is preserved.

But I & everyone overlooked the selection of *spontaneous variations* without which I think you have clearly shown that many of the phenomena are insoluble.

You have shown that the doctrine furnished explanations to phenomena otherwise inexplicable.[44]

Much attention has been given to determining whether or not Social Darwinism is an accurate designation for the immense formations and techniques that informed the era of the World Wars. Just as the establishment of natural selection as the primary mechanism would have been possible even if Darwin had been lost on the voyage of the *Beagle*, so too would the prejudices of Social Darwinism have existed, if only under the different and vastly more appropriate names of Spencerism, Sumnerism or monism.[45] The wide variety of appropriations of Darwinian theory, from Kropotkin and Nesmyth to Haeckel and Charles Murray, has always been noted and so comes as no surprise given that the 'transvaluation of all biological values' that Darwin accomplished left much uncertainty.[46]

The phrase 'survival of the fittest' did not appear in *The Origin of Species* until the 1869 edition. In a July 1866 letter, Wallace suggested that Darwin stop using the phrase natural selection in favour of Spencer's phrase 'survival of the fittest'. Wallace wrote that '[t]he term "survival of the fittest" is the plain expression of the fact; "natural selection" is a metaphorical expression of it, and to a certain degree indirect and incorrect, since ... nature ... does not so much select special varieties as exterminate the most unfavorable ones'. Darwin replied that Wallace's letter had arrived too late for him to alter the latest edition, but he promised to use 'survival of the fittest' in his next work, *Domestication of Animals*, and in any future editions of *The Origin of Species*. Wallace objected to 'natural selection' because Darwin's use of the term always carried with it a double meaning: '1st, for the simple reservation of favorable and rejection of unfavorable variations, in which case it is equivalent to the survival of the fittest, and 2ndly, for the *effect* or *change* produced by this preservation'.[47]

Darwin replied further that both Spencer and the popular French philosopher Paul Janet misuse or misunderstand the term natural selection.[48] Darwin wrote of his belief that the term natural selection 'was of great advantage to bring into connection natural and artificial selection'. This led him to use 'a phrase in common with Spencer' and despite Spencer's 'continually using the words, natural selection' Darwin continued to believe that the phrase 'had some advantage' because it had become quite common and so, despite his promise to use 'survival of the fittest' in the future, he wrote 'Natural Selection has now been so largely used abroad and at home, that I doubt whether it could be given up, and with all its faults I should be sorry to see the attempt made. Whether it will be rejected must now depend on the survival of the fittest'.[49] Wallace was the only one to mention to Darwin his double use of natural selection, and Darwin wrote to him that in his view this 'blunder has done no harm'. While he might have overemphasized 'favorable variation' in his work, Darwin replied that Wallace had himself a times 'put the opposite side too strongly'. Darwin agreed to use the phrase survival of the fittest, but he would consistently couple it with natural selection and referred to it as a metaphorical expression, just as Wallace accused him of using natural selection. Subtly, as with Malthus, Darwin reinterprets survival of the fittest not as 'a plain expression of fact', but as 'a metaphor for *effect* and *change*'. It was in relation to Malthusian thinking, cruelty and slavery that Darwin came to touch upon the question of self-interest:

> It is argued that self-interest will prevent excessive cruelty; as if self-interest protected our domestic animals, which are far less likely than degraded slaves, to stir up the rage of their masters. It is an argument long since protested against with noble feeling, and strikingly exemplified, by the ever illustrious Humboldt. It is often attempted to palliate slavery by comparing the state of slaves with our poorer countrymen: if the misery of our poor be caused not by the laws of nature, but by our institutions, great is our sin; but how this bears on slavery, I cannot see; as well might the use of the thumb-screw be defended in one land, by showing that men in another land suffered from some dreadful disease.[50]

Darwin's repeated use of the phrase 'battle for life' is also interesting in relation to Spencer's phrases 'struggle for existence' and 'survival of the fittest'.[51] They do not all mean the same thing, and the phrases 'battle for life' and 'recurrent struggle for life' appear in Darwin's works as early as the 1830s. It must be remembered that Darwin was not a theorist, nor was he a natural philosopher. He was an observer and collector who developed perhaps the most important theory for the establishment of the sciences of life. So it is always possible when analysing Darwin's choice of words to over-interpret, but as a careful observer and noting the care with which he qualified such phrases as 'struggle for existence' as metaphors, certainly it is not going too far to wonder if this delay in the use of 'survival of the fittest' was a careful and deliberate choice; a choice open to the same person who chose not to mention the word evolution until the last word of the last sentence of *The Origin of Species*. The battle for life is the work of achieving and maintaining life in the face of the equally strong and inevitably, at the level of the particular individual, victorious death. The struggle for existence is the struggle to determine who or which groups will carry out this battle for life. These will be the fittest. The results of the battle for life cover the globe, just as it 'fills with its dead and broken branches the dust of the earth'. The struggle for life is the mutualistic relation of all species, while the struggle for existence is between individuals of the same species.⁵² It is the struggle for life that is implied by the title of *The Origin of Species*.

Individuals, even of the same species, differ in small ways from each other, variation results from the preservation of some of these small changes, or differences, 'profitable to an individual' in its 'infinitely complex relations to other organic beings' and its relations to 'external nature'. He says later that 'the plant at the edge of the desert is said to struggle for life against the drought', or 'when we reach the Arctic regions, or snow-capped summits, or absolute deserts, the struggle for life is almost exclusively with the elements'.⁵³ The current understanding of these relations remained slim in Darwin's time, and he was

> convinced of our ignorance of the mutual relations of all organic beings ... Throw up a handful of feathers, and all must fall to the ground according to definite laws; but how simple is this problem compared to the action and reaction of the innumerable plants and animals which have determined, in the course of centuries, the proportional numbers and kinds of trees now growing on the old Indian ruins!⁵⁴

The struggle for life is not constant, as one might expect if he were following the doctrine that nature is in a perpetual state of war, but occurs 'at some period' of life or 'some season of the year'. The struggle might occur during each generation, or at intervals, but each species must

> struggle for life, and to suffer great destruction. When we reflect on this struggle [i.e., between populations rather than between individuals] for life, we may console ourselves with the full belief that the war of nature is not incessant, that no fear is felt, that death is generally prompt, and that the vigorous, the healthy, and the happy survive and multiply.⁵⁵

A theory of crisis rests comfortable amidst Darwin's spontaneity and gradualism. The struggle for existence comes at a certain time, a moment of crisis. Natural selection is the principle by which small and useful variations are preserved and it works within a natural world filled with crises, chance events and desolate regions. Here, too, we get the sense that Malthus's work was only a theoretical appropriation and a call to popular authority. If left to itself, any species would fill the space of the world to which it is adapted, but it does not. '[L]ook at a plant in the midst of its range, why does it not double or quadruple its number? We know it can perfectly well withstand a little more heat or cold, dampness or dryness, for elsewhere it ranges into slightly hotter or colder, damper or drier districts.' This complex relationship of the species to its environment is illustrated in its most simple form by the predator/prey relation.

> In the case of every species, many different checks, acting at different periods of life, and during different seasons or years, probably come into play; some one check or some few being generally the most potent, but all concurring in determining the average number or even the existence of the species. In some cases it can be shown that

widely-different checks act on the same species in different districts. When we look at the plants and bushes clothing an entangled bank, we are tempted to attribute their proportional numbers and kinds to what we call chance. But how false a view is this! Every one has heard that when an American forest is cut down, a very different vegetation springs up; but it has been observed that the trees now growing on the ancient Indian mounds, in the Southern United States, display the same beautiful diversity and proportion of kinds as in the surrounding virgin forests. What a struggle between the several kinds of trees must here have gone on during long centuries, each annually scattering its seeds by the thousand; what war between insect and insect – between insects, snails, and other animals with birds and beasts of prey – all striving to increase, and all feeding on each other or on the trees or their seeds and seedlings, or on the other plants which first clothed the ground and thus checked the growth of the trees![56]

Darwin reconciled gradualism with sudden events or crises in his use of the metaphors of the struggle for life and the struggle for existence. How to understand crises in nature and in society was a important question for Ricardo and Marx, to mention just two, but it was also to be found in Cuvier's theories of revolution and extinction. In Darwin, gradualism seems to be the view when he is discussing the possibility of crises on a large scale – as in geology or in the development of civilized mortality – but immediate and potentially catastrophic crises occur on the particular level of the individual. Darwin's gradualism has been amended in the present day by Nils Eldredge and Stephen Jay Gould's theory of punctuated equilibrium,[57] but in the context of Darwin's time, gradualism was indispensable if he was to give evolution a revaluation. In the late 1850s, any other argument would have left open a return to preformism, catastrophes or design.

Instinct and Slave-Making

The Aristotelian relation of master and slave, in all its naturalism and fixity, was generally accepted as authoritative through the ages – usually to justify some quite specific application of terror or tyranny. After many long detours and sudden dead ends, the good soul of the master became the fitness of the individual. The good body became the civilized body, or at least the body and soul worth cultivating and civilizing. The comparison of the habits of social animals had brought many naturalists to the view that instincts were themselves evidence of design, and the paternalistic relations of domination and subordination inherent in the master and slave were therefore believed to be perfectly logical justifications for slavery in the Americas.

Darwin used three examples in his chapter on instinct to argue that even the most seemingly 'miraculous' and 'wonderful' instinct could have come about by means of natural selection: the nesting behaviour of birds, the hive-building behaviour of bees and wasps and the slave-making behaviour of some ants. Although the first two are important to Darwin's argument, it is the discussion of slave-making behaviour that is central to the chapter.[58] It is worth pausing to note that although *The Origin of Species* is an intervention into the species question as a rational justification of slavery, Darwin made no mention of slavery until the chapter titled 'Instinct'. What better place than a chapter on exactly those traits that many assumed could not be changed by human intervention: domination as a special, 'wonderful', creation of the designer? Had not Aristotle made the master/slave relation a fundamental condition of life, and perhaps most importantly of *human* nature, too? Not only should animals be separated into master and slave, but also into those that live solitary lives and those that 'are disposed to combine for social purposes'. These latter gregarious animals can be classed as social animals. To be considered social animals, all must come together with 'one common objective in view ... Such social creatures are man, the bee, the wasp, the ant, and the crane.' Any differences in the social systems of these quite different species are smoothed out by the continuity of the authoritarian impulse that underlies them all. Some 'submit to a ruler, others are subject to no governance: as, for instance, the crane and the several sorts of bee submit to a ruler, whereas ants and numerous other creatures are every one his own master'.[59]

How are we to make sense of the existence of slave-making ants? Do ants and humans embody the instinct for authority, venturing out only to gather new slaves to work in their colonies and living on the fruits of slave-labour? Since only certain species are enslaved, does their victimization embody an impulse to submission and domination? If slavery and mastery are essential aspects of nature, one would be inclined to believe in slavery as an essential relationship in nature. If this is truly an act of special creation, Darwin asks, then why does the behaviour vary so much across the gregarious species? And if it does vary within ant, or for that matter human, societies, then can slave-making be a special creation? Variation in the behaviour led Darwin to ask how such an instinct came about, and the connection between ants and other gregarious slave-makers was later made clear in quite materialist language in *The Descent of Man*: 'The brain of an ant', he writes, 'is one of the most marvellous atoms of matter in the world, perhaps more so than the brain of a man'.[60]

Rather than beginning with contemporary human slave-making and tracing back the slave-making instinct, Darwin begins and ends with criticizing slavery as a 'most wonderful' specimen of design and quite characteristically relies on his own observations. Of particular interest was any geographic variation in the social structure of slave-making colonies. Jean-Pierre Huber had discovered two species of slave-making ants, *Formica rufrscens* and *Formica sanguine*, published to a wide audience in the United States in his *The Natural History of Ants*. Thoreau mentions this work in his famous description of the battle of the ants in *Walden*.[61] Slave-making amongst ants was a controversial subject since it served to support slavery as a natural condition.[62]

According to Huber, *F. rufescens* is completely dependent on slave-making to support its colonies. 'The species would certainly become extinct in a single year without its slaves. The workers or sterile females, though most energetic and courageous in capturing slaves, do no other work.' Darwin's description is not one that flatters the slave-makers, whose general behaviour Darwin views as having been deeply altered by their peculiar habit.

> They are incapable of making their own nests, or feeding their own larvae. When the old nest is found inconvenient, and they have to migrate, it is the slaves which determine the migration, and actually carry the masters in their jaws. So utterly helpless are the masters, that when Huber shut up thirty of them without a slave, but with plenty of food which they liked best, and with their larvae and pupae to stimulate their work, they did nothing; they could not even feed themselves, and many perished of hunger.[63]

If *F. rufescens* was the only species of slave-making ant, it would seem to be strong evidence of special creation, for 'What can be more extraordinary than these well ascertained facts?' One species that can do nothing more than rule and war, and another that lives to be dominated and is happiest when it is tending to the master's needs. 'If we had not known of any other slave-making ant, it would have been hopeless to have speculated how so wonderful an instinct could have been perfected.'[64] But there are other slave-making ants, and Darwin does not simply speculate because one, *Formica sanguinae*, was by chance available to his direct observation in 1857, twelve years after Thoreau witnessed his great battle at Walden Pond. Darwin's 'wonderful' and 'perfect' finds, are not quite the appropriate words to describe this instinct except with his own deep irony.

Darwin praised Huber's powers of observation as being beyond reproach and had relied heavily on Huber's description as *F. sanguinea* was rare in England, and Huber's specimens were taken in Switzerland. Darwin's correspondence with Frederick Smith, entomologist of the British Museum, is of great importance here, and much of it finds its way into the chapter on instinct. The two continued to correspond and occasionally visit until Smith's death in 1879.[65] Smith's main publications were on the ants of the British Isles, and Darwin wrote that while he has no quarrel with the work of either naturalist, he was convinced that the works of Huber and Smith were fundamentally at odds. Huber claimed that *F. sanguinea* behaviour is fixed, that it does not work except to go on slaving raids. However, Smith observed quite different and much more varied behaviour in the colonies of *F. sanguinea* he had located in England, where it is a rare species confined to a limited geographic range. Such variation in so 'wonderful' and 'perfect' an instinct was curious. Why should there be any variation in a divine creation? This question, he goes on to argue, is even more important given that Huber discovered a second slave-making ant, one which Darwin had the opportunity to observe directly over the course of the three years leading up to the publication of *The Origin of Species*. The discussion of slave-making ants did not appear in either the 1842 or 1844 preliminary drafts of the chapter on instinct. Darwin, always in poor health after his return from the voyage of the *Beagle*, had penned the essays with instructions and funds for their publication should he die before the project was done. In these preliminary sketches, only the second, longer, version has an extensive discussion of instinct, and ants are not discussed at all. Darwin's observations on slave-making ants are certainly late additions to *The Origin of Species*, and it would be easy to overstate the importance of this discussion to the overall work, but it is possible to trace the connection between this late addition and the arguments concerning slavery as a natural social relation and a social relation justified by nature itself.

The specialized bodies of ants and their complex social instinct presented Darwin with 'the acme of difficulty on our principle of natural selection'.[66] Darwin asked Smith to provide him with any 'remarkable instances of disparity in form, etc. – in workers or Insects livings in community'.[67] The reason behind this request is quite clear: bodies that appeared specially created for servitude were, of course, not only an old Aristotelian assumption, but conclusive evidence supporting the dominant polygenic theories of human origin. Smith furnished Darwin with an example from Brazil, which he sketched in 'relative proportion' and sent along with the specimen and an invitation to meet whenever Darwin visited the Museum, where Smith was the specialist on social insects. Their correspondence suggests that Darwin visited Smith on occasion in the following months. A February 1858 letter from Smith to Darwin confirmed that 'your ant with the bright red head is as you suspect the *F. sanguinea* – the other is *F. Rufa*'. Smith knew of Darwin's great interest in observing the colonies, and warned him that February is 'too early for predatory attack' and that the frequency of such attacks varies with the kind and number of larvae and pupae present. Those times when the most worker larvae are present in the slave colony are the periods of greatest predation by *F. sanguinea*. According to Smith, *F. sanguinea* reduces the number of raids when there is an abundance of queens and males in the colonies they attack, since these are only good as food sources and of no use as slaves.[68] Smith not only advised Darwin an ants, but on the nest-building behaviour of bees and wasps as well. Darwin responded to Smith's identification of his specimen and advice with a letter of 9 March 1858, asking four questions which bear directly upon his interest in instinct and especially the social and slave-making instincts.

Darwin inquired first about Smith's general observations of 'stray workers inhabiting the nest of a distinct species'. Had Smith ever seen fertile bees in the nest of another species, or were there only sterile workers? This question was the only one concerning bees or the other social insects that Darwin asked. The other questions all bear directly on slave-making ants. Smith wrote his answers on Darwin's questionnaire and returned it to him, and so fortunately both questions and answers are preserved in Darwin's correspondence. To the first question, Smith responded simply and definitively with the single word 'Never'. Likewise, Darwin asked if Smith had ever observed stray workers in the nests of ants that do not make slaves. Smith wrote 'Not stray workers – there are communities of *Myrmica nitidula* & *M. muscorum* that live in harmony with *Formica rufa*. I never saw stray workers in the nests of *non-slavemaking* species' (original emphasis). Darwin wanted to know to what degree the slave-making behaviour varies, if it does in fact vary within and across species. Does *F. sanguinea*, he asked, 'invariably' take slaves, 'or is it only an occasional yet frequent instinct?' Smith replied that he had never observed *F. sanguinea* 'without numbers of other species' and that 'no one has recorded any different habit'. Darwin later interpreted this to mean that there is variation in the numbers of slave-making ants present and in the species of ants that are enslaved in individual colonies, though Smith actually believed that the usual proportion of slaves to masters remains fairly constant at 'about eight in twenty'. Finally, Darwin asked about the direct contradiction between the accounts of Huber and Smith regarding the variation in the behaviour of the slave-making ants. 'Is it typical', he asked, for either of the species that are enslaved by *F. furescens* to return to their own nests to feed their queens and males, as with those species that Smith had found to 'live in harmony', or do these queens and males feed themselves? Smith answered that the slaves had never been observed to enter or leave the nest and so Smith 'is induced' to believe that the slaves' main tasks are to move the egg brood 'as the temperature requires and also in enlarging the nest, etc'. In contrast to Huber, Smith wrote that he had observed *F. sanguinea* workers supply all of the food for the colony. 'I have seen hundreds of them on young larch trees attending on a black swollen Aphis'.[69] He expanded on this point in another letter to Darwin a month later. Smith noted that his most thorough observations of *F. sanguinea* were made during the period from early May to the middle of August. He explained that during this peak period, 'the greater number of slaves I have found more than once' in August. Smith repeated that he had never

> observed any Slave-Ant either issue from or enter the nest – I have seen them repeatedly *carried in* in their perfect condition – and once in the larvae pupae state. The workers regularly come and go as the temperature varies during the day, virtually disappearing underground when the sun shone its greatest intensity, and emerging again in the cooler afternoon.

Smith continued with this characterization of the slave ants: 'I have been induced from my own observations to consider them – household-Slaves – which perform some drudgery in the nest – You are at perfect liberty to quote the results of my imperfect observations'.[70] Darwin did use them, relying much more on Smith than Huber for the discussion in *The Origin of Species.*

Darwin was quite excited by his stumbling upon the slave-making *F. sanguinea* and the enslaved *F. rufa* when one of his health crises necessitated a trip to Moor Park for hydropathy, a water cure. Darwin wrote to his son 'I pass my time [at the spa] chiefly in watching the ants, & I find that though many thousands inhabit each hillock, each seems to know all its comrades, for they pitch unmercifully into a stranger brought from another ant-hill'. It was there that he discovered colonies of *F. sanguinea* that he could directly observe. 'Yesterday I was poorly', he wrote, 'but I got better in the evening & am very well today. I can not walk far yet; but I loiter for hours in the park & amuse myself by watching the Ants: I have great hopes I have found the rare Slave-making species & have sent a specimen to Brit[ish] Mus[eum] to know whether it is so'.[71] The specimen was sent to and identified by Smith. Upon his return to Down House, Darwin wrote to Thomas Hooker that 'As usual Hydropathy has made a man of me for a short time', but even more important is that 'I had such a piece of Luck at Moor Park. I found the rare slave-making Ant, & saw the little black niggers [*F. nigra*] in their Master's nest'.[72]

One is often now reminded of Wallace's near simultaneous working out of the theory of natural selection, but it is good to recall, when one asks why Darwin delayed so long the publication of *The Origin of Species*, his illness and pain, his taking of water cures and observing of slave-making ants on the eve of the American Civil War. Then, too, one might remember Thoreau and how his own observation of ants was joined to the events surrounding the peculiar institution. Thoreau decided to finish the record of his observations in the manner of earlier naturalists, and so noted that the epic battle of the ants 'which I witnessed, took place in the Presidency of Polk, five years before the passage of Webster's Fugitive Slave Bill'.[73]

Darwin undertook his observations of ants with great intensity, and his conclusions on the variation of their behaviour gave evidence against the fixity of species and against slavery by design. How indeed did so 'wonderful an instinct' come to exist? One clue is Darwin's use of the term 'wonderful', which is often used by him when referring to evidence supporting fixity – evidence that he is just about to stand on its head and put to the service of variation and descent. It 'would be hopeless to speculate' without the variation in the behaviour of the slave-makers and it is this variation that offers the missing piece to the puzzle.

Darwin's observations given in *The Origin of Species* generally follow the line of questioning he had put to Smith over the previous three years. Darwin recounts that he 'opened fourteen nests of *F. sanguinea* and found a few slaves in all', just as Smith had observed. As to the presence of fertile females and males amongst the slaves, they were only found in their own colonies, and never in the nest of *F. sanguinea*. Smith had said that he had never observed the slaves enter or leave, and Darwin notes that it is easy to distinguish the slave-making ants as 'the slaves are black and not above half the size of their red masters'. Wanting to see for himself whether the slaves leave the colony, Darwin disturbed several nests and found that

> when the nest is disturbed, the slaves occasionally come out, and like their masters are much agitated and defend the nest: when the nest is much disturbed and the larvae and pupae are exposed, the slaves work energetically with their masters in carrying them away to a place of safety.

Smith, never having seen the slaves outside the nest, referred to them as household slaves who perform drudgery tasks, but Darwin implies that there might be more to their behaviour in the nest: 'It is clear that the slaves feel quite at home' and have no need of 'an overseer or a whip-man'. Rather than doing all the work while their masters reclined or raided for more slaves, Darwin observed 'for many hours several nests in Surrey and Sussex, and never saw a slave either enter or leave ... The masters, on the other hand, may be constantly seen bringing in materials for the nest, and food of all kinds'. Darwin notes that his opportunities to observe *F. sanguinea* occurred over the course of three years in Surrey, but only during the months of June and July when he suspected, and Smith's letter confirmed, that there would be fewer slaves present than later in August. However, Smith's observations covered the months of May, June and August in Surrey and Hampshire. Between the two, their observations covered the most active periods for the *F. sanguinea* colonies. Neither Darwin nor Smith observed the enslaved *F. rufa* leave the nest. However, observations made 'during the current year', i.e., of the publication of *The Origin of Species*, produced evidence of variation in the behaviour of the ants from what Huber had described. 'In the month of July' – a month for which Smith had no observations:

> I came across a community with an unusually large stock of slaves, and I observed a few slaves mingled with their masters leaving the nest, and marching along the same rock to a tall Scotch-fir tree, twenty-five yards distant, which they ascended together probably in search of aphides or cocci.

Now Darwin had two pieces of evidence for variation in this 'most wonderful' behaviour and most difficult challenge to his theory of natural selection. He could now demonstrate that variation exists between species of slave-making ants, and in addition, he could point to variations in behaviour within a species brought about by changes in the conditions of life: the availability of slaves and the quantity and type of food available to the colony. Gently, Darwin contrasted his observations with Huber's to show that slave-making could have resulted from basic food foraging. He again praised Huber's powers of observation and the consistency of his observations that the slave masters 'alone open and close the doors in the morning and evening; and, as Huber expressly states, their principal office is to search for aphids'. Darwin observed that 'this difference in the usual habits of the masters and slaves in the two countries probably depends merely on the slaves being captured in greater numbers in Switzerland than in England'.[74]

Darwin did not doubt Huber's observations, but he was not convinced that slave-making behaviour is fixed. Huber certainly saw an ordered commonwealth when he described a colony:

> In regarding those colonies, which exist at our very feet, and where so much harmony and order prevail, I think, I perceive the Author of nature, tracing with his all-powerful hand, the laws of a republic exempt from abuse, or framing the model of those compound [i.e., slave-holding] societies, where servitude is allied to a common interest.

Huber discussed and classified the types of ants in terms of their morphological features and their social structure. Huber began his text by discussing the study of ants in the context of the study of government, a study facilitated by classification and a common nomenclature. Indeed, the advances in classification made it possible for the naturalist to devote attention to questions of government, habits, and industry. His preface begins:

> Much has been written upon Ants: their form of government, and their labours, excited the admiration of the ancients equally with the moderns; but it is only in the present day that just observations have taken the place of the fabulous recitals of Pliny and Aristotle.
>
> The Naturalists of the last century attended to their transformations, discovered the sexes, and cleared up many essential points of their history. Learned anatomists, also, described their organs, classed the different kinds of Ants, and pointed out their generic characters.
>
> To the individual who wishes to be acquainted with the history of these insects, it is of no small advantage, to be enabled to designate the species, without lying under the necessity of entering into long and minute descriptions; he can devote himself entirely to the study of those laws by which these various tribes are governed, undertake new researches into their habitudes and industry, and have his attention solely occupied with the phenomena their instinct presents ...
>
> The History of Ants being yet incomplete, I have been induced to join my own observations to those of my learned predecessors, trusting that the perseverance with which I have studied the economy of Ants for several years has enabled me to fill up a portion of that void which still remains in this branch of science.[75]

Darwin witnessed a migration of *F. sanguinea* 'and it was a most interesting spectacle to behold the masters carrying as Huber has described, their slaves in their jaws'. In *F. rufescens*, Huber described the slaves carrying the masters, but no matter. During his taking of the water-cure, Darwin observed the end of a slave-making raid by a colony of *F. sanguinea* on a colony of *F. fusca*. There too, the masters carried their new slave larvae and pupae, as well as the bodies of the *F. fusca* that had died defending their nest. Darwin traced 'the returning file for about forty yards, to a very thick clump of heath' but was unable to find the nest of the raided colony. The last of the raiders was just emerging 'from the heath carrying a pupa', when he arrived. The forty yards was as much a scientific observation as it was a comment on Darwin's own physical health. He notes that the raid, like any slave raid, was a battle of the ants. In the wake of the combat, he finds that 'two or three individuals of the *F. fusca* were rushing about in the greatest agitation, and one was perched motionless with its own pupa in its mouth on top of a spray of heath over its ravaged home'.[76]

Darwin witnessed another combat where he described the slave-makers as 'haunting' the spot where he had witnessed the earlier raid. These 'tyrants' were 'evidently not in search of food. They approached and were vigorously repulsed by an independent community of the slave species (*F. fusca*); sometimes as many as three of these ants clinging to the legs of the slave-making *F. sanguinea*'. The slave-makers 'ruthlessly killed their small opponents and carried their dead bodies as food to their nest; twenty-nine yards distant; but they were prevented from getting any pupae to rear as slaves'. *F. fusca* was perhaps not made for servitude after all, for if so, why would they resist so violently and only submit if they have been raised as slaves in the nest of *F. sanguinea*? To test his hypothesis, Darwin placed a number of *F. fusca* pupae 'on a bare spot near the place of combat', where were 'eagerly seized and carried off by the tyrants who perhaps fancied that, after all, they had been victorious in their late combat'. Darwin then placed the pupae of 'the little and furious' *F. flava*, which is rarely found enslaved, on the same battlefield. The *F. sanguinea* that had eagerly taken the *F. fusca* pupae 'were terrified when they came across the pupae, or even the earth from the nest of *F. flava*, and quickly ran away'. They returned when 'all the little yellow ants had crawled away' and the slavers only then 'took heart and carried off the pupae ... Such are the facts, though they did not need confirmation by me, in regard to the wonderful instinct of making slaves'.[77] The contradictions between Huber, Smith and himself are the result of the differences in the behaviour of the ants themselves, and not differences in the observers. The instinct to make slaves might be wonderful, but it is not wondrous. One can contrast the behaviour of *F. sanguinea* with that of *F. rufescans* in terms of nest-building, migration patterns, food collection and rearing of young: *F. rufescans* 'cannot even feed itself, it is absolutely dependent on its numerous slaves'. But there were also differences between the *F. sanguinea* observed in Switzerland by Huber and those in England observed by Smith and Darwin.

> In Switzerland the slaves and masters work together, making and bringing materials for the nest: both, but chiefly the slaves, tend, and milk as it may be called, their aphids; thus both collect food for the community. In England the masters alone usu-

ally leave the nest to collect building supplies and food for themselves, their slaves, and larvae. So that the masters in this country receive much less service from their slaves than they do in Switzerland.

The explanation for the 'wonderful' instinct is not so wondrous, but simply natural selection, variation, and chances:

> By what steps the instinct of *F. sanguinea* originated I will not pretend to conjecture. But as ants, which are not slave-makers, will, as I have seen, carry off pupae of other species, if scattered near their nests, it is possible that pupae originally stored as food might become developed; and the ants thus unintentionally reared would then follow their proper instincts, and do what work they could. If their presence proved useful to the species which had seized them – if it were more advantageous to this species to capture workers than to procreate them – the habit of collecting pupae originally for food might by natural selection be strengthened and rendered permanent for the very different purpose of raising slaves. When the instinct was once acquired, if carried out to a much less extent even than in our British *F. sanguinea*, which, as we have seen, is less aided by its slaves than the same species in Switzerland, I can see no difficulty in natural selection increasing and modifying the instinct – always supposing each modification to be of use to the species – until an ant was formed as abjectly dependent on its slaves as is the *Formica rufescens*.

Is the implication that human slavery could be the result of the carrying back of enemies for sacrifice and cannibalism? Perhaps such a terrible origin would be poetic, but it is clear from the variation of slavery that it is not a fixed condition, for even this instinct varies:

> I have endeavoured briefly in this chapter to show that the mental qualities of our domestic animals vary, and that the variations are inherited. Still more briefly I have attempted to show that instincts vary slightly in a state of nature ... I can see no difficulty, under changing conditions of life, in natural selection accumulating slight modifications of instinct to any extent, in any useful direction. In some cases habit or use and disuse have probably come into play. I do not pretend that the facts given in this chapter strengthen in any great degree my theory; but none of the cases of difficulty, to the best of my judgment, annihilate it. On the other hand, the fact that instincts are not always absolutely perfect and are liable to mistakes; – that no instinct has been produced for the exclusive good of other animals, but that each animal takes advantage of the instincts of others; – that the canon in natural history, of 'natura non facit saltum' is applicable to instincts as well as to corporeal structure, and is plainly explicable on the foregoing views, but is otherwise inexplicable, – all tend to corroborate the theory of natural selection.[78]

The masters are dependent on slaves but not the slaves on the masters. It is the slave-making species that is altered by slavery and not the victims of the tyrants. The slave-making behaviour of ants could be explained by selection, and could not be used to justify human slavery. Nor was the master/slaver relation as wonderful or wondrous as it might have seemed. An essential intellectual support for slavery had been removed and science could stand as much against slavery as it had previously stood with it. As a gradualist, small differences were always important in Darwin's thinking, and to have found variation in the behaviour of *F. sanguinea* was the grain of sand on the balance.

The Origin of Species asks the central question of Darwin's time: what explains the origins and variety of species? The existence of variation was obvious to any observer, as Darwin notes at the beginning of his work. As early as 1850, a reviewer of the works of Prichard, Morton, Blumenbach, W. Lawrence and Samuel Stanlope Smith was led to begin his essay by noting that '[i]n surveying the globe in reference to the different appearances of mankind, the most extraordinary diversities are apparent to the most superficial observer ... Hence arises the question – *Have all these diverse races descended from a single stock?*'[79] Human variety held the key to the species question precisely because the question always referred to human variety, and because Linné, Cuvier and Lamarck had the wisdom to place humans in the animal kingdom. Variation in one could explain variation in all, since the process working on one is the same process working on all.

But Darwin was not a recapitulist. The struggle for life and the struggle for existence are not the same, so the evidence does not point to the recapitulation of phylogeny and ontogeny, but to a fundamentally genealogical commonality. Humans and apes are connected, Linné had said, in a great chain of being that embodies the movement towards perfection. Haeckel and Spencer added the movement of increasing complexity as the essential to construct the biogenetic law and a law of cosmic evolution. To the contrary, Darwin's theory was neither optimistic nor teleological: genealogy rather than being connected all things in nature. Linné did not believe in extinction but that the forms at the bottom of the chain of being were constantly spontaneously generated. Darwin notes that, however fanciful, spontaneous generation was the only materialist or scientific explanation that ran counter to creationism – at least until the theory of natural selection.[80]

> It is, no doubt, extremely difficult even to conjecture by what gradations many structures have been perfected, more especially amongst broken and failing groups of organic beings; but we see so many strange gradations in nature, as is proclaimed by the canon, 'Natura non facit saltum,' that we ought to be extremely cautious in saying that any organ or instinct, or any whole being, could not have arrived at its present state by many graduated steps. There are, it must be admitted, cases of special difficulty on the theory of natural selection; and one of the most curious of these is the existence of two or three defined castes of workers or sterile females in the same community of ants; but I have attempted to show how this difficulty can be mastered.[81]

Darwin executes more than just a rhetorical reversal with the naming of *The Origin of Species*. Realizing that the 'the question of the origin of species is of the human species',[82] Darwin chose to avoid the question of human origins because to do so would have been to play on his opponents' board and make his work a part of the monogenic/polygenic debate. In breaking with that discursive formation Darwin overturns the species question by demanding that we consider humans to be just one of an infinite variety of species, all of which were created by the same processes that could even now be seen at work. Darwin shifts man from a central place in understanding the variety of nature and so produces the break from the old debate. If humans can tell us so much about the origins of the vast cacophony of nature, then there was no reason to privilege humans as the special key to knowledge. Any species could answer some or all of the question of origins. The scientific ideology which had been key to classification and to polygenic theory – that the answer lay in analogy, being, philology, comparative anatomy and recapitulation – was turned on its head by Darwin's account of a single common line of descent shaped by natural selection, among other conditions of life. Darwin does not directly refer to polygenism until ten years later in the *Descent of Man*, and by then the polygenists had already been eclipsed by the combined forces of Darwin's critique and the American Civil War.

Darwin combined the genealogical classification of species with a theory of population fluctuations and the gradual accumulation of small variations – 'a grain of sand is enough to tip the balance' in terms of survival. Darwin's work is profoundly materialistic; it relies upon the continuity of genealogy. The natural world moves from being a closed system capable of being rendered in tables of classification towards an an open field where 'from so simple a beginning endless forms most beautiful and most wonderful have been, and are being evolved'.[83] This is famously the only instance in *The Origin of Species* where the word evolution is used, and it is significant that it is used in this passage to juxtapose the fixity of the law of gravity with the plasticity of descent with modification and the workings of natural selection.

Most simply put, Darwin made the question of human origins a matter of the origin of any species. Humans were no longer at the centre of the study of life and the natural history of any species could now explain the origin of human beings. Linné may have placed man in the chart of classification and as the measure and explanation for its origins, and Agassiz repeated the medieval representation of man as the crown of creation in his chart. Darwin, however, placed humans in a genealogical tree of life, that is, directly in nature itself, and allowed that other species shall now explain the origin of man. *The Origin of Species* opens us to the infinity of nature and makes humans just one of many species joined in life's great struggle to exist 'whilst this planet has gone cycling according to the fixed law of gravity'.[84]

Life and its struggles to continue occupied the centre of scientific knowledge, and the displacement of man that Darwin brought about could not be tolerated by natural history. Biology and ecology would now begin to supplant natural history as sociology and economics would now replace political economy. In the time of the present book – which was begun four years before the war on terror and six years before the passing of the Patriot Act, as Thoreau might have dated it – natural history is today no longer a discipline of knowledge but a style of nature writing consumed by those who consider it entertaining to read about someone else's experience of nature. The end of natural history came with the end of the dispute between the monogenists and polygenists.

From *The Voyage of the Beagle* to *The Descent of Man*: The End of the Monogenesis/Polygenesis Discourse

As was noted, the only instance in *The Origin of Species* where Darwin uses the word 'evolved' is in the context of his juxtaposing the fixity of the law of gravity with the plasticity of his theory of descent with modification, a plasticity due in large part to the chance workings of Natural Selection. In the argument that precedes its first and only use, Darwin was at pains to ensure that his views regarding evolution were not confused with those of his contemporaries. Evolution, as in the unfolding of nature, had become established as a common philosophical/scientific concept and moral belief. Gould reminded us of several reasons for Darwin's careful hesitation at the mention of evolution. For one, the word had already been appropriated by embryologists in a very precise and circumscribed sense in relation to the homunculus, the adult within the germ cell: In 1744, when the German biologist Albrecht von Haller coined the term, he chose carefully, for the Latin *evovere* means 'to unroll'; indeed, the tiny homunculus unfolded from its originally cramped quarters and simply increased in size during its embryonic development.[85] Such a preformist theory was incompatible with Darwin's understanding of variation. If every individual or species was already there from the beginning, then descent with modification would be, if not unnecessary, then certainly impossible.

It was not evolution in its preformist meaning that provided continuity in nature, instead natural selection supplied an understanding of the discontinuity of nature without recourse to fanciful theories of successive creations. Descent with modification combined genealogical classification with a definition of evolution that was opposed to preformism. In countering the theory of the fixity of species, descent with modification provided a seemingly infinite number of possibilities for reading the history of nature in the history of species.[86]

A well-recognized difficulty of the theory was Darwin's acknowledged failure to understand the means for the transmission of hereditary traits. Mendel was only then delivering the results of his experiments, which were to remain obscure and virtually unread for another fifty years. Darwin himself did not grasp the significance of Mendel's paper. Darwin had experimented with peas during the same time, though not with the same hypothesis in mind, as his interest in hybrids was in their fertility, rather than their tendency to vary from generation to generation.[87] Consistent with his materialist tendencies, Darwin adopted a Lamarckian theory of paragenesis, and along with it the possibility of social reform and progress that later writers would apply for good and ill. It also broke with later theorists who, like August Weismann, would posit an unchanging and eternal 'germ-line'. Darwin's 'germules' provided the possibility that environment – the conditions of life – could influence the transmission of characteristics from one generation to the next. Weismann, in contrast, posited a germ-line 'thoroughly insulated from influences emanating from the body'. Darwin's germules would 'make characteristics acquired by the body transmissible to the next generation, while the biospores [of Weismann] could not'.[88] The immutability of the germ-plasm would necessitate its own set of social relations and itself become a true end of history, for 'the germ-plasm can not be formed anew, it can only grow, multiply, and be transmitted from one generation to another'.[89]

The lack of a hereditary mechanism did not deter Darwin from publishing to the degree that the slave-making instinct did. The level of difficulty he encountered was not at all the same. Whatever the specific mechanism, all of the observable facts and accumulated knowledge confirmed the genealogical relation of life. The implications of this were clear: no animal is superior to another in some grand plan of creation or chain of being. Scientific observation and not religious dogma had now established that no variety of human is superior to another and races are not distinct and hierarchically arranged. As Darwin's friend Asa Gray would put it years later with much subtlety, 'Races, in fact, can hardly, if at all be said to exist independently of man'. Differences exist on a continuum which they fall along by chance and are preserved by selection. 'But man does not produce [races]. Such peculiarities – often surprising enough – now and then originate, we know not how (the plant *sports*, as the gardeners say); They are only preserved propagated and generally further developed, by the cultivators care.'[90] No longer could slavery look to scientific rationality for support, though it did not take more than a few months after the end of the American Civil War for Thomas Huxley to hold forth that slavery was an evil primarily because of its effects on the slaveholder. In reality, Negroes really were inferior and should be treated as such, Huxley concluded, just as women are maturally inferior to men. While slavery could no longer organize scientific knowledge, the Negro, the criminal, the woman, the degenerate and the deviant would over the coming century find equal opportunities in the sciences of life and society.[91]

Despite his Victorian manners we know that Darwin was deeply opposed to slavery: 'I would not be a Tory, if it was merely on account of their hard hearts about the scandal to Christian nations – slavery'.[92] Although repulsed by the living conditions and manners of the Feugians, he had sailed with three who were returning after years of assimilating English norms. Darwin claims to have been unaware of them until the arrival of the *Beagle* at Tierra del Feugo. The three were Fitz-Roy's charges and he was returning them to their homes after their having been kidnapped and brought back to be civilized during the first voyage of the *Beagle*. Darwin knew the Feugian dubbed Jemmy Button well enough to write that he 'believed in dreams, though not ... in the devil'. Their homecoming was disastrous for all concerned.[93]

The voyage of the *Beagle* transformed Darwin from a believer in special creation – one of Fitz-Roy's many requirements in a bunkmate – into the now familiar originator of the theory of natural selection. Darwin wrote of being mocked but also respected by the worldly sailors on the *Beagle*, who often referred to him as 'professor' as much for his conventional religious views and moralism as for his intellectual pursuits. He was at the beginning of the voyage led by

> inward conviction of the existence of God, and of the immortality of the soul ... the state of mind which grand scenes formerly excited in me, and which was intimately connected with a belief in God, do not essentially differ from that which is often called a sense of sublimity, and however difficult it may be to explain the genesis of the sense, it can hardly be advanced as an argument for the existence of God any more than the powerful though vague and similar feelings excited by music.[94]

Unlike his view of nature, his views on slavery did not change as a result of the expedition. Darwin was before and remained after his return an ardent opponent of slavery. 'I was told before leaving England that after living in slave countries all my opinions would be altered; the only alteration I am aware of is forming a much higher estimate of the negro character.'[95] Darwin understood the potential political and social implication of his work no less than did the polygenists. He, too, sought to free scientific inquiry from dogma, just as the polygenists insisted. Darwin understood that scientific knowledge had effects on social relations, and believed that over time human society could be transformed by scientific knowledge. This transformation might not come through a frontal assault, nor in an instant, but rather by creating the conditions for the accumulation of new knowledges. Biology and sociology are two results of these newly accumulated strata.

As a stylist, Darwin can be simultaneously gentle and damning. His comments on Thomas Carlyle are particularly representative of his technique of criticism.

> Carlyle sneered at almost everyone ... His expression was that of a depressed, almost despondent yet benevolent man; and it is notorious how heartily he laughed. I believe that his benevolence was real, though stained by not a little jealousy. No one can doubt about his extraordinary power of drawing pictures of things and men – far

more vivid, as it appears to me, than any drawn by Macaulay. Whether his pictures of
men were true ones is another question …

It is astonishing to me that Kingsley should have spoken of him as a man well
fitted to advance science … As far as I could judge, I never met a man with a mind so
ill adapted for scientific research.[96]

Darwin did not write to engage in day-to-day politics, but nonetheless his writings were interventions in the politics of his day. He modestly noted that even
his 'few sentences [on slavery] were merely an explosion of feeling'.[97] One hears
this in his further comment on Carlyle:

He has been all-powerful in impressing some grand moral truths on the minds of
men. On the other hand, his views about slavery were revolting. In his eyes might
was right. His mind seemed to me a very narrow one; even if all branches of science,
which he despised, are excluded.[98]

Almost twenty years after he first described his explosion of feeling, it had not
subsided. In June 1861, he wrote to Gray, 'Great God. How I should like to see
the greatest curse on earth – slavery – abolished'.[99] According to Francis Darwin,
his father was haunted all his life by his experiences in the slave territories and
the nightmarish scenes frequently returned to him in his dreams.[100]

To understand the relation of Darwin's work to slavery, there is perhaps no
better place to begin than Darwin's relationship with Fitz-Roy and his experiences while aboard the *Beagle*. Darwin was grateful to Fitz-Roy for taking him
on the voyage and was careful in his public criticism of one who had given him
so much. As he wrote of Fitz-Roy in the preface to *The Voyage of the Beagle*, 'the
opportunities, which I enjoyed of studying the Natural History of the different countries we visited, have been wholly due to Captain Fitz-Roy … during
the five years we were together, I received from him the most cordial friendship and steady assistance'.[101] There is little hint given of their frequent disputes
and numerous difficulties. Even when writing privately Darwin blamed his difficulties with Fitz-Roy on the captain's seemingly Shakespearean tragic flaws.[102]
'I have always felt that I owe to the voyage the first real training or education of
my mind; I was led to attend closely to several branches of natural history, and
thus my powers of observation were improved, though they were always fairly
developed'. Despite his 'notable features' and a disposition that was 'generous
to a fault', Darwin wrote that Fitz-Roy was prone to mood swings, that he was
obsessive and 'He seemed to me often to fail in sound judgment or common-
sense. He was very kind to me, but was a man very difficult to live with on the
intimate terms which necessarily followed from our messing by ourselves in the
same cabin'. Like any British naval commander of the time, he was a contributor
to the ruthless discipline for which the Royal Navy was noted. 'Fitz-Roy's tem-
per was a most unfortunate one … [it was] usually worst in the early morning,

and with his eagle eye he could generally detect something amiss about the ship, and was then unsparing in his blame.'[103]

For his part, Fitz-Roy upon meeting Darwin immediately sized up his worthiness for the demanding voyage. He would later take pride in his having selected a crew that withstood the five-year voyage with only a few deaths. That he was a good judge of character is not really a matter to be disputed, nor are his considerable skills as a naval captain and as a naturalist in doubt. What is of interest is the method he used. Examining Darwin's facial profile, Fitz-Roy declared that the shape of Darwin's nose indicated his being unlikely to survive the voyage. Darwin relates that Fitz-Roy informed him that he had almost been rejected

> on account of the shape of my nose! Fitz-Roy was a disciple of [the phrenologist] Lavater, and was convinced that he could judge a man's character by the outline of his features, and he doubted whether any one with my nose could possess sufficient energy and determination for the voyage. But I think he was afterwards well satisfied that my nose had spoken falsely ... The voyage of the *Beagle* has been by far the most important event in my life and has determined my whole career, and yet depended on ... such a trifle as the shape of my nose.[104]

It is not difficult to understand that Darwin would be grateful to Fitz-Roy. However, Darwin also admitted that their relationship was often strained. Not only did the arrangements of their quarters, Fitz-Roy's phrenological passion and Darwin's too frequent seasickness create difficulties, but so too did Fitz-Roy's ardent support of slavery. Darwin embarked on the voyage a true believer in creation and a firm opponent of slavery. Like Agassiz, his actual experiences of Africans were limited at best, though the shipyards and docks of England through which he passed did not lack diversity. Darwin had been assured by many that his anti-slavery views would be altered by contact with the various others. On the contrary, and while on the voyage, he wrote to his sister and friends that just the opposite had happened. 'I have seen enough of Slavery and the dispositions of the Negroes, to be thoroughly disgusted with the lies and nonsense one hears on the subjects in England'.[105]

The conflicts between Darwin and Fitz-Roy go unmentioned in the early published account of the voyage. Fitz-Roy's own position on slavery was in reality a reformist one. His affinity for phrenology, belief in the great chain of being, varied scientific interests – the space he shared with Darwin was crammed with twenty-two chronometers – and Christian charity gave him what was at the time a common view that slavery was a necessary evil – ultimately good for the slaves despite the burdens on their masters. Not surprisingly, slavery skewed the paternal function in a manner that a captain in the Royal Navy could appreciate and that a Christian reformer could embrace. The Fuegians had been his special project, and the death of one caused Fitz-Roy to remark on how little distance there was really between them. Transformed into Europeans, he returned them to their former homes where they were to serve as the kind of civilizing influences that would in due time bring about an end of the slave trade, which Fitz-Roy and other reformers believed would be halted through the administration of the tropical areas by those indigenous people that had been raised, like the Fuegians, to a semblance of European respectability.

> The chief, if not the only cause of the slave trade in Brazil, is want of population – want of an industrious population, able as well as willing to clear away primeval forests, and render the soil fit for culture – able to work in the open fields under a hot sun, to cultivate the sugar cane, cotton plants, mandioca, and other productions of tropical climates.
>
> While this extensive and most powerful cause exists, selfish, unprincipled owners of immense territories in Brazil, and elsewhere, will not refrain from importing hundreds, even thousands of unhappy wretches, who, once landed, become the helpless instruments of immense gain to their owners: neither can any reasonable number of shipping efficiently blockade the coasts of two great continents.
>
> If I am right in these assertions, it appears that there is no method by which the slave trade can be totally suppressed, except by destroying the cause of so abominable a traffic: and that, to this end, a native population should be encouraged in hot climates, who, being gradually inured to work on their native soil, for remuneration from their employers, and a prospect of future comfort for themselves and their offspring, would totally supersede the demand for constrained labour. Of course, the only way by which such a result could be obtained – I should say, perhaps, the first step towards so satisfactory a result, would be, that the government of a slave-importing country should declare that trade piratical: and proclaim every human being free; bound to no man, free to do any thing not contrary to religion, or law, from the moment he or she embarked on board a vessel belonging to that country, or placed a foot upon its soil; which might then indeed be termed, in common with our happy land, a sacred soil. By such a plan as this, individuals would suffer for a time, but the mass of society would be gainers incalculably.[106]

The tropics might be theirs alone, but European culture was the right of all. It would of necessity come in its own good time and so bring an end to what Fitz-Roy himself said was the abominable slave trade.

Fitz-Roy's impressions of Bahia were much in keeping with the civilizing impulse of some moderates and apologists for slavery. Bahia was a country rich in natural resources, and in Fitz-Roy's view had thrived under Portuguese rule. Once the rule of the colonial power was withdrawn, Bahia became overrun with such troubles that frightened away potential foreign investment. Fitz-Roy described the approach to Bahia in picturesque terms:

> As we sailed in rapidly from the monotonous sea, and passed close along the steep but luxuriantly wooded north shore, we were much struck by the pleasing view. After the light-house was passed, those by whom the scene was unexpected were agreeably

surprised by a mass of wood, clinging to a steep bank, which rose abruptly from the dark-blue sea, showing every tint of green, enlivened by bright sunshine, and contrasted by deep shadow: and the general charm was heightened by turreted churches and convents, whose white walls appeared above the waving palm trees; by numerous shipping at anchor or under sail; by the delicate airy sails of innumerable canoes; and by the city itself, rising like an amphitheatre from the water-side to the crest of the heights.

For Fitz-Roy, Bahia was a confirmation of his views on slavery and possibilities for social progress and regression.

Bahia has declined ever since its separation from Portugal: unsettled, weak governments, occupied too constantly by party strife to be able to attend to the real improvement of their country, have successively misruled it. Revolutions, and risings of the negro population, interrupting trade, have repeatedly harassed that rich and beautiful country, and are still impending.

Were property secure, and industry encouraged, the trade from Bahia might be very extensive, particularly in sugar and cotton: but who will embark much capital upon so insecure a foundation as is there offered?

The immense extent and increase of the slave population is an evil long foreseen and now severely felt. Humanely as the Brazilians in general treat their slaves, no one can suppose that any benevolence will eradicate feelings excited by the situation of those human beings. Hitherto the obstacles to combinations and general revolt among the negroes, have been ignorance, mutual distrust, and the fact of their being natives of various countries, speaking different languages, and in many cases hostile to each other, to a degree that hardly their hatred of white men can cause them to conquer, even for their immediate advantage.

The slave trade has already entailed some of its lamentable consequences upon the Brazilians, in demoralizing them by extreme indolence, and its sure accompaniment, gross sensuality; but there are in store afflictions hitherto unfelt, occasioned by the growing hordes of enemies who are yearly causing more perplexity and dread in the territories of Brazil.

Could the Brazilians see clearly their own position, unanimously condemn and prevent the selfish conduct of individuals, emancipate the slaves now in their country, and decidedly prevent the introduction of more, Brazil would commence a career of prosperity, and her population would increase in an unlimited degree. In that immense and most fertile country, distress cannot be caused by numerous inhabitants; food is abundant, and the slight clothing required in so warm a climate is easily procured.[107]

John Bachman and Fitz-Roy would certainly have had an interesting and mutually supportive conversation on such matters of slavery and civilization.

Darwin gives this account of an argument he had with Fitz-Roy over the captain's defence and praise of slavery 'early in the voyage at Bahia'. Darwin wrote that when he told Fitz-Roy that he 'abominated' slavery, the captain

told me that he had just visited a great slave owner, who had called up many of his slaves and asked them whether they were happy, and whether they wished to be free, and all answered 'No.' I then asked him, perhaps with a sneer, whether he thought that the answer of slaves in the presence of their master was worth anything? This made him excessively angry, and he said that as I doubted his word we could not live any longer together. I thought that I should have been compelled to leave the ship; but as soon as the news spread, which it did quickly, as the captain sent for the first lieutenant to assuage his anger by abusing me, I was deeply gratified by receiving an invitation from all the gun-room officers to mess with them. But after a few hours Fitz-Roy showed his usual magnanimity by sending an officer to me with an apology and a request that I would continue to live with him.[108]

Fitz-Roy's name may not be commonly known today, but he is indirectly remembered everywhere. He was the founder of the British Meteorological Society and established a network of weather observatories linked by telegraph that could warn of gales and provide other weather data. He issued the first weather 'forecast', a term he coined, on 8 August 1861. During his tenure as governor of New Zealand he raised the ire of the European settlers by his more reasonable and tolerant treatment of the Maoris than any of his predecessors. Fitz-Roy had left on the voyage of the *Beagle* favouring the gradualist geology of Lyell, but later strongly defined the notion of biblical creation. Imagine the confusion of one believing in biblical authority and determinism while creating a meteorological forecasting network, the chaotic nature of which must have both exhilarated and frustrated one so dedicated to prediction. The fact that Darwin only refers to his apology for slavery indicates the depth of Darwin's opposition, rather than the degree of Fitz-Roy's support. Fitz-Roy writes of his great desire to 'civilize' the captive Feugians, of his sadness at the death of one. According to his biographers, Fitz-Roy's correspondence indicates the great attention he paid to their well-being on the voyage and his shock at the reaction of the Feugians upon the return of the captives. He, of course, seems to make little connection between the events which led to their seizure and transfer to England and the seizure of others.

Their infamous meeting at Oxford was one event that Darwin regretted, and after Admiral Fitz-Roy donned his uniform for the last time and retired to his rooms to slit his own throat in 1864, Darwin contributed to the fund for Fitz-Roy's family.[109] The reasons for his suicide are unclear and those close to him had little warning of it. We do know that Fitz-Roy's work and travels had taken their toll on him, as he was also known for his reticence. For example, Darwin was stunned to learn upon their return to England that Fitz-Roy had been engaged the entire time of the voyage, marrying almost as soon as he disembarked. For five years, Fitz-Roy had not mentioned his fiancée at all to his cabin-mate.

During the voyage, Darwin would frequently note the horrific effects of slavery in his letters and his journals, from the severing of familial relationships – which must have concerned one apt to think in terms of descent and genealogy – to its very existence as the 'great stain' on humanity. Darwin's views gathered strength during his journey through the slave countries. In his *Journal of Researches*, Darwin relates the *Beagle* passing 'under one of the massive, bare, and steep hills of granite which are so common to this country'. At the top, he learned, was a small area now 'notorious' as having been the site of a village of runaway slaves who had managed to avoid the authorities for many years. Once discovered, soldiers were of course immediately dispatched and 'the whole were seized with the exception of one old woman, who, sooner than again be led into slavery, dashed herself to pieces from the summit of the mountain'. Darwin remarks bitterly that such defiance 'in a Roman matron' would have been taken as evidence of 'the noble love of freedom [but] in a poor negress' it is hypocritically dismissed as 'mere brutal obstinacy'.[110]

In three letters – one written to his sister and the others to friends – penned only eleven days apart, Darwin wrote of his commitment to abolitionism.

> Hurrah for the honest Whigs! I trust they will soon attack that monstrous stain on our boasted liberty, Colonial Slavery. I have seen enough of Slavery and the dispositions of the Negroes, to be thoroughly disgusted with the lies and nonsense one hears on the subject in England. Thank God, the cold-hearted Tories, who, as J. Mackintosh used to say, have no enthusiasm, except against enthusiasm, have for the present run their race.[111]

Darwin suggested to his sister that 'some of the anti-slavery people' they knew should investigate the activities of a British agent in Rio 'who has a large salary to prevent the landing of slaves; he lives in Botofogo, and yet that was the bay where, during my residence, the greater number of smuggled slaves were landed'.[112] From Rio, Darwin wrote to J. S. Henslow that 'I would not be a Tory, if it was merely on account of their cold hearts about that scandal to Christian nations – Slavery'.[113]

A small digression is in order to point out this passage from Darwin's son Francis which relates Darwin's reactions to slavery and cruelty:

> It was in November 1875 that my father gave his evidence before the Royal Commission on Vivisection ... Something has already been said of my father's strong feeling with regard to suffering both in man and beast. It was indeed one of the strongest feelings in his nature, and was exemplified in matters small and great, in his sympathy with the educational miseries of dancing dogs, or in his horror at the sufferings of slaves. (He once made an attempt to free a patient in a mad-house, who (as he wrongly supposed) was sane. He had some correspondence with the gardener at the asylum, and on one occasion he found a letter from a patient enclosed with one from the gardener. The letter was rational in tone and declared that the writer was sane and wrongfully confined.
>
> My father wrote to the Lunacy Commissioners (without explaining the source of his information) and in due time heard that the man had been visited by the Commissioners, and that he was certainly insane. Sometime afterwards the patient was discharged, and wrote to thank my father for his interference, adding that he had undoubtedly been insane, when he wrote his former letter.)

The remembrance of screams, or other sounds heard in Brazil, when he was powerless to interfere with what he believed to be the torture of a slave, haunted him for years, especially at night. In smaller matters, where he could interfere, he did so vigorously. He returned one day from his walk pale and faint from having seen a horse ill-used, and from the agitation of violently remonstrating with the man. On another occasion he saw a horse-breaker teaching his son to ride, the little boy was frightened and the man was rough; my father stopped, and jumping out of the carriage reproved the man in no measured terms.

One other little incident may be mentioned, showing that his humanity to animals was well-known in his own neighbourhood. A visitor, driving from Orpington to Down, told the man to go faster, 'Why,'said the driver, 'If I had whipped the horse THIS much, driving Mr. Darwin, he would have got out of the carriage and abused me well.'

With respect to the special point under consideration, – the sufferings of animals subjected to experiment, – nothing could show a stronger feeling than the following extract from a letter to Professor Ray Lankester (March 22, 1871):–

'You ask about my opinion on vivisection. I quite agree that it is justifiable for real investigations on physiology; but not for mere damnable and detestable curiosity. It is a subject which makes me sick with horror, so I will not say another word about it, else I shall not sleep to-night.'[114]

Thirteen years after his letter to Henslow, slavery would still easily elicit from Darwin what he called 'an explosion of feeling'. Even his close relationship with Lyell did not dull his sharp criticism of any seeming support for slavery.

I was delighted with your letter in which you touch on Slavery; I wish the same feelings had been apparent in your published discussion. But I will not write on this subject, I should perhaps annoy you, and most certainly myself. I have exhaled myself with a paragraph or two in my *Journal* on the sin of Brazilian slavery; you perhaps will think that it is in answer to you; but such is not the case. I have remarked on nothing which I did not hear on the coast of South America. My few sentences, however, are merely an explosion of feeling.

Darwin's patience had been greatly tested by the letter from Lyell. He misread Lyell's letter in which he was really relating the views of a planter perhaps in an attempt to produce a response from Darwin; which it did.

How could you relate so placidly that atrocious sentiment (in the passage referred to, Lyell does not give his own views, but those of a planter.) about separating children from their parents; and in the next page speak of being distressed at the whites not having prospered; I assure you the contrast made me exclaim out. But I have broken my intention, and so no more on this odious deadly subject.[115]

Darwin consistently found the argument concerning the civilizing effects of slavery to be one of the most difficult to rebut. Appealing to the self-interest of the master in avoiding his own degeneration from the application of authority was useless. Having established slavery as variable, the variability in the peculiar institution was itself a difficulty, for surely just as one might find a brutal and greedy master, one might also find a benevolent patrician hard at the work of civilization. Darwin drew a stark contrast between two such slavers, and amidst the contrast he observed as well that both were capable of the same extreme cruelty. Darwin's description of Socego and its neighbouring estates seems to take into account a difference without ever accepting the existence of slavery in either its patriarchal or brutal manifestations.

Entertained by the gracious owner of the estate at Socego, Darwin was taken with the abundance of food, the well kept and orderly arrangement of buildings, the training of the slaves in 'Various trades' and the generally bucolic atmosphere that could have only stirred the imagination of one schooled on the stories of imperial patriarchy and the polis. Here under the lordship of Manuel Figuireda, the father-in-law of one of Darwin's overland travelling companions, stood a quadrangle formed by the master's house, store houses and workshops, in the middle which lay coffee piled high for drying in the sun. Around these were arranged 'the huts of about 110 negroes whom Señor Figuireda and one white man as manager contrive to keep in perfect order'. These buildings stood on a little hill, overlooking the cultivated ground, and surrounded on every side by a wall of dark green luxuriant forest'.[116] In his diary written at the time, he described the site simply as 'a piece of cleared ground cut out of the almost boundless forest'.[117] The land was so fertile and game so abundant that at their meals 'if the tables did not groan, the guests surely did'. Señor Figuireda allowed his slaves Saturday and Sundays to 'work for themselves, and in this fertile climate the layout of two days is sufficient to support a man and his family for the whole week'. Rising early, Darwin was especially taken by the way in which the 'solemn stillness' of the dawn was 'broken by the morning hymn, raised on high by the whole body of blacks' in their morning call to work.[118]

As though taking up his argument with Fitz-Roy, Darwin admitted that he was sure that the slaves on this estate 'pass happy and contented lives' and thus there was 'something exceedingly fascinating in this simple and patriarchal style of living: it was such a perfect retirement and independence from the world' that the appearance of a stranger such as himself was announced to the 'rocks and the woods but to nothing else'.[119] But it was a 'simple and patriarchal style' that Darwin could not ultimately accept. In his diary he described his host as a 'villain' but then crossed that out and inserted the phrase an 'enterprising character' who had 'cut excellent roads, built industry, and brought modern medicine to the region'.[120]

Just as the gilded chairs and sofas of the master's sitting room 'oddly contrasted with the whitewash walls, thatched roof, and windows without glass', of the 'simple and uncomfortable house',[121] so too was the simplicity and abundance of Figuireda's estate at odds with the real relations upon which it thrived.

It was an 'exceedingly fascinating' place only 'so long as the idea of slavery could be banished', which, of course, was impossible.[122] For every slaver like Figuireda, there were many more brutal masters. It is easy, Darwin observed, for the former to be transformed into the latter, as Darwin would soon observe during his own plunge into the heart of darkness. Independent from the rest of the world, except of course for the coffee, slave and luxury goods trades, the lure of patriarchal power was certainly strong at Figuireda's estate.

As Darwin journeyed along the Rio Macae towards 'the last patch of cultivated ground in that direction', he was more and more overwhelmed by the vastness of the jungle that 'abounded with beautiful objects' and 'the proportion of cultivated ground can scarcely be considered as anything; compared to that which is left in a state of nature'. Travelling along a road where 'instead of milestones, the roadside is marked with crosses, to signify where human blood has spilled' or so overgrown that it was necessary to cut a path, Darwin wrote that he 'enjoyed the never failing delight of riding through the forests'.[123] Darwin was headed for the estate of another of his travelling companions, a Mr Lennon. Stopping to rest for the night at a less developed fazenda a few miles from Lennon's, Darwin was not only taken in by the 'vast extent of forests' surrounding this estate, but also in the contrast between the treatment of the slaves here, who 'appeared miserably overworked & badly clothed. Long after dark they were employed.' Here, despite the same abundance was none of the tranquillity of Figuireda's fazenda, and here the slaves were not maintained 'by the common method' of giving them two days a week to themselves, but put to work every day as befitted their alternative instruction in the building of civilization.[124]

Events and circumstances turned even worse when they finally came to Lennon's estate – 'the most interior piece of cleared ground' – and Lennon immediately fell into a violent argument with his agent, a Mr Cowper. The argument, which was over unpaid debts, rapidly grew to such intensity that Darwin wrote later of wanting to leave immediately, but had really nowhere to go. In the later *Journal of the Researches*, he would write that here he was an eyewitness to 'one of those atrocious acts which can only take place in a slave country'.[125] In his diary, he seems hesitant to detail the argument, as he first mentions only that the fight was such that he wanted to flee. This brief mention is followed by some additional details about the size of Lennon's fazenda, the weather, and the state of his own health, as Darwin had several bouts of illness on the trip into the interior. Only then does he abruptly return to the circumstances of his arrival. In the *Diary*, Darwin relates the events in this way:

> During Mr. Lennon's quarrel with his agent, he threatened to sell at the public auction an illegitimate mulatto child to whom Mr. Cowper was much attached: also he nearly put into execution taking all the women & children from their husbands

and selling them separately at the market at Rio. Can two more horrible & flagrant instances be imagined?

But this monstrous act was not the act of a monster or driven by instinct, but of a common and good man.

> And yet I will pledge myself that in humanity and good feeling Mr. Lennon is above the common run of men. How strange and inexplicable is the effect of habit and interest! Against such fact how weak are the arguments of those who maintain that slavery is a tolerable evil.[126]

The account published in the *Journal of Researches* is richer in detail, but not in sentiment:

> Owing to a quarrel and a law-suit, the owner was on the point of taking all the women and children from the male slaves, and selling them separately at the public auction at Rio. Interest, and not any feeling of compassion, prevented this act. Indeed, I do not believe the inhumanity of separating thirty families, who had lived together for many years, even occurred to the owner. Yet I will pledge myself, that in humanity and good feeling he was superior to the common run of men. It may be said there exists no limit to the blindness of interest and selfish habit.[127]

It was agreed by all that Señor Figuireda was the person to arbitrate the dispute, and all began the next day to journey back to Socego. Once they arrived, Darwin immediately withdrew from his companions, who no doubt were focused on the dispute, and threw himself into his collecting and research. The judgement of Señor Figuireda and the result of the dispute goes unmentioned by Darwin. He does, however, repeat in the *Journal* the observation that he made in his *Diary* that day: leaving his companions to their dispute, he walked through the forest, with its 'numberless species' and 'grand scenes', of which 'it is impossible to give an adequate idea of the higher feelings which are excited; wonder, astonishment and sublime devotion filled and elevated the mind'. This he experienced in the midst of the horrors of slavery. 'These two days were spent at Socêgo, & was the most enjoyable part of the whole expedition; the greater part of them was spent in the woods, & I succeeded in collecting many insects & reptiles.'[128]

Over the next few days, Darwin and the remainder of the party retraced the route back to Rio de Janeiro and brought an end to their 'pleasant little excursion'.[129] He pauses in his account to reinforce his view that no one can avoid the barbarity induced by the social relations of slavery, for even if one is not a master, one remains in the position of a master. He knew this from his own experience, as he was shocked to find himself taken for a slave-master.

> I may mention one very trifling anecdote, which at the time struck me more forcibly than any story of cruelty. I was crossing a ferry with a negro, who was uncommonly stupid. In endeavouring to make him understand, I talked loud, and made signs, in

doing which I passed my hand near his face. He, I suppose, thought I was in a passion, and was going to strike him; for instantly, with a frightened look and half-shut eyes, he dropped his hands. I shall never forget my feelings of surprise, disgust, and shame, at seeing a great powerful man afraid even to ward off a blow, directed, as he thought, at his face. This man had been trained to a degradation lower than the slavery of the most helpless animal.[130]

Darwin took all of these events to heart when he rejected any notion of reformism or suggestion that the horrors of slavery could be mitigated by the self-interest of the master, who would be damaging a valuable article of wealth. It was exactly acting against his economic 'interest' that such a fine person as Mr Lennon threatened to tear apart the families of his plantation and sell them off to the highest bidder. Upon emerging from the Brazilian forest, Darwin denounced those who would excuse slavery by comparing the slave's condition to that of the industrial worker.

> It is often attempted to palliate slavery by comparing the state of slaves with our poorer countrymen: if the misery of our poor be caused not by the laws of nature, but by our institutions, great is our sin; but how this bears on slavery, I cannot see; as well might the use of the thumb-screw be defended in one land, by showing that men in another land suffered from some dreadful disease. Those who look tenderly at the slave-owner, and with a cold heart at the slave, never seem to put themselves into the position of the latter; – what a cheerless prospect, with not even a hope of change! picture to yourself the chance, ever hanging over you, of your wife and your little children – those objects which nature urges even the slave to call his own – being torn from you and sold like beasts to the first bidder! And these deeds are done and palliated by men, who profess to love their neighbours as themselves, who believe in God, and pray that his Will be done on earth! It makes one's blood boil, yet heart tremble, to think that we Englishmen and our American descendants, with their boastful cry of liberty, have been and are so guilty: but it is a consolation to reflect, that we at least have made a greater sacrifice, than ever made by any nation, to expiate our sin.[131]

With the advent of the Civil War, Darwin's enthusiasm for the Union and hopes that the war would bring about abolition and emancipation were tempered by his fear that a southern victory would bring about an expansion of slave territories and a permanent enshrinement of the peculiar institution. Like many in the first years of a war punctuated by the all-too frequent failure of the general staff of the Union to press its material advantages effectively, Darwin felt sure that the South could not be defeated militarily and worried that the United Kingdom would be brought into the war.

> I agree with much of what you say, and I hope to God we English are utterly wrong in doubting (1) whether the N. can conquer the S.; (2) whether the N. has many friends in the South, and (3) whether you noble men of Massachusetts are right in transferring your own good feelings to the men of Washington. Again I say I hope to God we are wrong in doubting on these points. It is number (3) which alone causes

England not to be enthusiastic with you. What it may be in Lancashire I know not, but in S. England cotton has nothing whatever to do with our doubts ... all I can say is that Massachusetts and the adjoining States have the full sympathy of every good man whom I see; and this sympathy would be extended to the whole Federal States, if we could be persuaded that your feelings were at all common to them.[132]

As much as the potential British intervention gave Darwin pause, so too did he fear that the movement for abolition in the empire might be lost if Great Britain was pulled into the American conflict. If this happened, the cause would be rolled back in both nations for perhaps generations. Before Gettysburg and Sherman's March to the Sea, Darwin most notably expressed his views on the conflict in letters to his friends Gray and Hooker. In 1861, he wrote to Hooker that while he had no interest or time to read Spencer, he did recommend Frederick Law Olmsted's *Journey in the Back Country*, 'for its an admirably lively picture of man and slavery in the Southern States'.[133] He mentions to Gray his having 'well studied' Olmsted's book along with John Elliott Cairnes's *Slave Power*, and of reading 'everything word of news' about the American conflict.

> I read Cairns's excellent Lecture, which shows so well how your quarrel arose from Slavery. It made me for a time wish honestly for the North; but I could never help, though I tried, all the time thinking how we should be bullied and forced into a war by you, when you were triumphant. But I do most truly think it dreadful that the South, with its accursed slavery, should triumph, and spread the evil. I think if I had power, which thank God, I have not, I would let you conquer the border States, and all west of the Mississippi, and then force you to acknowledge the cotton States. For do you not now begin to doubt whether you can conquer and hold them? I have inflicted a long tirade on you.[134]

It seemed realistic, if not ideal, to Darwin that slavery might only be contained within the southern states and in time eliminated through legal and parliamentary action.

> If abolition does follow with your victory, the whole world will look brighter in my eyes, and in many eyes. It would be a great gain even to stop the spread of slavery into the Territories; if that be possible without abolition, which I should have doubted. You ought not to wonder so much at England's coldness, when you recollect at the commencement of the war how many propositions were made to get things back to the old state with the old line of latitude[135]

He was not opposed to the carnage that was to take place as the end of slavery was worth almost any price.

> Some few, and I am one of them, even wish to God, though at the loss of millions of lives, that the North would proclaim a crusade against slavery. In the long-run, a million horrid deaths would be amply repaid in the cause of humanity. I never knew the newspapers so profoundly interesting. North America does not do England justice; I

have not seen or heard of a soul who is not with the North ... What wonderful times we live in! Massachusetts seems to show noble enthusiasm. Great God! How I should like to see the greatest curse on earth – slavery – abolished![136]

And finally, at the end of the war, Darwin wrote again to Gray about the surrender of the Confederate forces at Richmond. Relations between Britain and the United States were still on his mind:

> the grand news of Richmond has stirred me up to write. I congratulate you, & I can do this honestly, as my reason has always urged & ordered me to be a hearty good wisher for the north, though I could not do so enthusiastically, as I felt we were so hated by you. Well I suppose we shall all be proved utterly wrong who thought that you could not entirely subdue the South. One thing I have always thought that the destruction of Slavery would be well worth a dozen years war.[137]

In *The Voyage of the Beagle*, Darwin brought into the open all of the events he witnessed in the slave countries, but also revealed how he still continued to suffer from the horrors of what he had seen there. His son's statement that even decades later his father endured nightmares of Brazil has a more than adequate foundation in Darwin's own writings. Here is a writer who noted every detail, who centred his work upon his own observations and those of others, who even noted the sound of the sands near Rio Madre when trodden upon by his horse, but who at times leaves out details of his own experiences because the memory so easily enrages and horrifies him. The contrast between the Brazil of infinite tangled banks and the horrific land of slavery found its way into Darwin's later writings. Even if he could never leave behind the Brazil of his nightmares, he was glad to sail away, never to return.

> On the 19th of August we finally left the shores of Brazil. I thank God, I shall never again visit a slave-country. To this day, if I hear a distant scream, it recalls with painful vividness my feelings, when passing a house near Pernambuco, I heard the most pitiable moans, and could not but suspect that some poor slave was being tortured, yet knew that I was as powerless as a child even to remonstrate. I suspected that these moans were from a tortured slave, for I was told that this was the case in another instance. Near Rio de Janeiro I lived opposite to an old lady, who kept screws to crush the fingers of her female slaves. I have stayed in a house where a young household mulatto, daily and hourly, was reviled, beaten, and persecuted enough to break the spirit of the lowest animal. I have seen a little boy, six or seven years old, struck thrice with a horse-whip (before I could interfere) on his naked head, for having handed me a glass of water not quite clean; I saw his father tremble at a mere glance from his master's eye. These latter cruelties were witnessed by me in a Spanish colony, in which it has always been said, that slaves are better treated than by the Portuguese, English, or other European nations. I have seen at Rio de Janeiro a powerful negro afraid to ward off a blow directed, as he thought, at his face. I was present when a kind-hearted man was on the point of separating forever the men, women, and little children of a large number of families who had long lived together. I will not even allude to the many heart-sickening

atrocities which I authentically heard of; – nor would I have mentioned the above revolting details, had I not met with several people, so blinded by the constitutional gaiety of the negro as to speak of slavery as a tolerable evil. Such people have generally visited at the houses of the upper classes, where the domestic slaves are usually well treated, and they have not, like myself, lived amongst the lower classes. Such inquirers will ask slaves about their condition; they forget that the slave must indeed be dull, who does not calculate on the chance of his answer reaching his master's ears.[138]

Of Darwin 'we can safely infer that he would have been unwilling to support the marriage of racism and evolution'.[139] Natural and sexual selection and *The Descent of Man* destroyed the polygenesis theory. Darwin is often characterized as apolitical, but politics obviously has its place in any scientific ideology and the rebuttal of a scientific ideology cannot avoid being simultaneously a scientific as well as a political intervention. When a writer mistook Darwin's views as advocacy for the kind of supremacy that Sumner and Galton looked forward to, Darwin was quick to clarify his views in reference to slavery.

> Permit me to add a few other remarks. I believe your criticism is quite just about my deficient historic spirit, for I am aware of my ignorance in this line. 'In the historic spirit, however, Mr. Darwin must fairly be pronounced deficient. When, for instance, he speaks of the "great sin of slavery" having been general among primitive nations, he forgets that, though to hold a slave would be a sinful degradation to a European to-day, the practice of turning prisoners of war into slaves, instead of butchering them, was not a sin at all, but marked a decided improvement in human manners.') On the other hand, if you should ever be led to read again Chapter III., and especially Chapter V., I think you will find that I am not amenable to all your strictures; though I felt that I was walking on a path unknown to me and full of pitfalls; but I had the advantage of previous discussions by able men. I tried to say most emphatically that a great philosopher, law-giver, etc., did far more for the progress of mankind by his writings or his example than by leaving a numerous offspring. I have endeavored to show how the struggle for existence between tribe and tribe depends on an advance in the moral and intellectual qualities of the members, and not merely on their capacity of obtaining food. When I speak of the necessity of a struggle for existence in order that mankind should advance still higher in the scale, I do not refer to the MOST, but 'to the MORE highly gifted men' being successful in the battle for life; I referred to my supposition of the men in any country being divided into two equal bodies – viz., the more and the less highly gifted, and to the former on an average succeeding best.[140]

For Darwin, there could no accommodation for even moderate views on slavery, whether the institution was being carried out by ants or men. The lonely worker holding the last surviving pupae after a raid by the slavers points us back to Darwin's experiences on his trip inland from Rio de Janeiro. In human societies, slavery led even the kind-hearted to commit atrocities that contradict nature's lack of cruelty. At the end of his 'pleasant little excursion', Darwin split off from the feuding slave owners and threw himself into the 'boundless' forest. There, 'the higher feelings of wonder, astonishment, and devotion' stood in marked contrast to what he had just seen. This moment reverberates through *The Origin of Species* and most obviously in the chapter on instinct. Not only in its argument against the scientific support of slavery, but the very view of nature as a 'tangled bank' can be found in the description of the forest at Socego that Darwin wrote while Mr Lennon and Mr Cowper made their cases to Señor Figuireda:

> The woods are so thick & matted that I found it quite impossible to leave the path. – the greater number of trees, although so lofty, are not more than from 3 to 4 feet in circumference ... The contrast of the Palms amongst other trees never fails to give the scene a most truly tropical appearance: the forests here are ornamented by one of the most elegant, the Cabbage-Palm; with a stem so narrow, that with the two hands it may be clasped, it waves its most elegant head from 30 to 50 feet above the ground. – The soft part, from which the leaves spring, affords a most excellent vegetable. – The woody creepers, themselves covered by creepers, are of great thickness, varying from 1 to nearly 2 feet in circumference. – Many of the older trees present a most curious spectacle, being covered with tresses of a liana, which much resembles bundles of hay. – If the eye is turned from the world of foliage above, to the ground, it is attracted by the extreme elegance of the leaves of numberless species of Ferns & Mimosas. – Effect of walking on Mimosa. Thus it is easy to specify individual objects of admiration; but it is nearly impossible to give an adequate idea of the higher feelings which are excited; wonder, astonishment & sublime devotion fill & elevate the mind.[141]

When he writes later that there is 'grandeur in this view of life', can there be but little doubt as to what the other views of life might be? Genealogical descent from a common ancestor cut through the then current discourse on European supremacy.

> Whether primeval man, when he possessed but a few arts, and those of the rudest kind, and when his power of language was extremely imperfect, would have deserved to be called man, must depend on the definition which we employ. In a series of forms graduating insensibly from some ape-like creature to man as he now exists, it would be impossible to fix on any definite point where the term 'man' ought to be used. But this is a matter of very little importance. So again, it is almost a matter of indifference whether the so-called races of man are thus designated, or are ranked as species or subspecies; but the latter term appears the more appropriate. Finally, we may conclude that when the principle of evolution is generally accepted, as it surely will be before long, the dispute between the monogenists and the polygenists will die a silent and unobserved death.[142]

Can we not see in this 'silent and unobserved death' the same death of man that a hundred years later Foucault discerned approaching? To do away with the dominant scientific ideology on race and human origins and to replace it with a monogenic materialist theory was Darwin's often unacknowledged achievement. The debates over the work of the American School of polygenesis formed the structure around which almost all discourses on race and human origin were ordered. In opposition to this, Darwin linked the process of adaptation (Cuvier was concerned primarily with morphology) with location, with habitat (just as species are not fixed, so too are habitats not fixed in time or space) and extinction. In such a dynamic and indeterminate system of nature, genealogical analysis gave to Darwin the only sure foundation for classification and therefore, for any scientific study of life:

> As all the organic beings, extinct and recent, which have ever lived on this earth have to be classed together, and as all have been connected by the finest gradations, the best, or indeed, if our collections were nearly perfect, the only possible arrangement, would be genealogical. Descent being on my view the hidden bond of connection which naturalists have been seeking under the term of the natural system.[143]

CONCLUSION: THE AUTHORITY OF THE SCIENCES OF LIFE

The history of scientific classifications is obviously not the history of human variety. The history of scientific classifications of human variety is the history of how those schemes contributed to the scientific ideology of race. It is via this route that the classification or system of classifications of human variety contributed mightily to the formulation of the concept of race. It is not that science alone gave us our concept of race. The history of the scientific classifications of human variety demonstrates the extent to which scientific classification is a dynamic – sometime cooperative, sometimes contradictory – technology of power. In the nineteenth century, scientific classification was but one of many contributors to racialist and segregationist ideology. There was, of course, much more going on in the world, and certainly the American South's peculiar institution would have made race a concern without any help from or reference to scientific ideology.

And yet, the scientific classification of human variety did serve to legitimize the disciplinary formations which emerged during the nineteenth century: the sciences of life, especially biology and ecology; and the sciences of society, especially sociology, social work and criminology, emerged from the wreckage of polygenism, natural history and political economy. These provided nodal points around which the scientific ideology of race could be formed and manipulated. Once this occurred, the everyday notion of race as the explanation for human variety could become the scientific view, and vice versa.

It is often assumed that sociology and biology arose in opposition to each other (if one excluded the organicists), but the preceding pages argue strongly that they arose together, each drawing on the other for mutual authority. In fact, the organicists prove this point by their very existence.

> Historians want to write histories of biology in the eighteenth century; but they do not realize that biology did not exist then, and that the pattern of knowledge that has been familiar to us for a hundred and fifty years is not valid for a previous period. And that, if biology was unknown, there was a simple reason for it: that life itself did not exist. All that existed was living being, which were viewed through a grid of knowledge constituted by Natural History.[1]

The definition of a science is not a problem under the ordinary circumstances of the authority of science. The definition of science is a problem only if we admit that science is social. Science is, after all, at its most basic level simply what scientists do and how, as a consequence, they are regarded by others. This is the portion of their work that enters into public discourse. Much of it – laboratory notes, field notes, formal and informal discussions with other scientists, correspondence, distractions, factual errors, etc. – never enter into 'public' discourse. The problem is that the very definition of science and of the scientist changes – and so too does the relationship of the scientist to science. This is easily demonstrated if one considers how different natural history with its curios and gardens is from the modern specialized laboratory and field scientists of today. One way to reduce this difficulty would be to focus on the mutual dependence of the sciences of life and society for theoretical and organizational schemes for ordering human existence.

Before one can enter into a discussion of the meaning and uses of race, one must first discuss science, for race does not precede science, rather, science first established race as we understand it today, but it did so for different purposes. Race is a system of classifications produced first and foremost to legitimate the ability of sociology and biology – the foremost 'life science' – to speak the truth and therefore enter into the realm of scientificity. So if histories have so much trouble tracing the origins of race through definition of the word race, it is because its origin does not lie in some far off time, but in our present era. Race as we think of it was not thought of in the same way during past eras. Race emerges with Enlightenment, in the European encounter with itself and with the other, and with its own systems of classification and discourse.[2] Is there a single language, epistemology or practice that signified race, that provided a historical continuity even when modified in different eras, i.e., despite obvious discontinuities, or is this another fruitless search for an origin in a past that is itself only the product of the concerns of today? This is why philology and etymologies are useless, because they examine themselves and so cannot reveal their own repressed histories without overcoming and eliminating themselves. If the genealogy of economic concepts and the genealogy of morals could be accomplished because change had become a part of history, Darwin faced having to make genealogy an aspect of analysis when the 'history' in natural history had always been a fixed and unmoving one, save for the late addition of catastrophes to account for fossil finds and geological observations. The American School was driven to uncover the truth of human variety in the freeing of science from religion. They too failed to reconcile fixity with variety, but they did pursue what they thought to be the logical conclusion of rational consideration of the species question. They found instead that the Jeffersonian impulse they had rejected had appeared again in the work of Darwin, the last of the bourgeois radicals, to wipe them from memory of science. 'If we consider all the races of man as forming a single species, his range is enormous ... the variability of man may with more truth be compared with that of widely-ranging species, than with that of domesticated animals.'[3]

To understand the origin of the truth of race in the United States is to understand it as the production of a widely dispersed discourse on life and society. If one traces the formation and transformations of this bio-social discourse as it appears as a scientific ideology one is led to speculate the following:

1. There are scientific ideologies of nature (biology, medicine, natural history) and their systems of classification, especially classifications of types and symptoms. We have seen how the problem of classification had its origins in the classification of human variety. We have also seen that there are significant discontinuities, contradictions, modifications, and lacunae in the history of these classifications. The work of Aristotle is not the same as that of Xenophon. Pliny differs from both of them and inaugurates an entirely different discourse on human variety. The research on the varieties of human established the priorities of coming to understand variety and of describing the monstrous in the very definition of human.

Darwin made it impossible to separate humans into different kingdoms or into different species. It was unfortunately not impossible to do just this division in terms of civilizations or community, however. 'The proliferation of living forms was not the only type of evidence on which the fact of evolutionary transformation was based. There was confirmatory evidence both from the study of vestiges and from the process of recapitulation.'[4] It has always seemed strange that American sociology embraced degeneration and Spencerism/social Darwinism, but it is not so hard to understand if one considers these other routes: the study of vestiges or atavist traits and the general acceptance of recapitulation as a guiding biological principle. The study of vestiges, which Lumbers proposed to be the avatars of criminality, coincided with the formation of sociology. One might draw the scheme of the sciences of man as moving from crainiology → polygenism → criminal anthropology and biometrics → physical anthropology and eugenics → sociology/criminology/forensics and social problems. The contradiction between an emancipatory multiplicity and the anxiety about the monstrous types would inevitably result from such a permanent production of difference by 'permanent variety-making machinery'.[5] The elaboration of variety in a system of organization is simultaneously the repression of it. Linnean classification was a repression of variety by the attempt to capture variety within the confines of a fixed table. Darwin supports the infinity variety of nature within the limitations of descent and modification, natural selection and chance.

2. There are scientific ideologies concerning the forces of life, both the rational forces (those which are themselves allied with the Enlightenment universals of reason, history and consciousness) as well as the irrational forces, e.g.,

the instincts, the opposing conceptions of the id in Weismann and Freud's use of the concept,[6] the mob, the mass, atavism) and also rationalized irrationality in the examples of the concepts of the market and the anarchy of the social relations of capital. Race does not begin in secularism but in opposition to religion. If we look to the period before Darwin, we can see that there were two problems that were really one and the same: what accounts for the variation we find in nature and what accounts for the variation we find in humans?

The question of the origin of this difference began as a theological one and ended as a scientific one. Race became a tool with which to combat religion because it had to rely on a secular mode of knowledge along with a secular view of the force of nature being life and not God. This concern with mortality is one that is sometimes lost in discussions on the origins of scientific classifications, for the concern with the differences that identify types manifested itself in the attempt to provide the religious authorities with a means to distinguish the speaker of truth from the heretic – or to force the heretic finally to speak the truth. There is a certain continuity between pseudo-Aristotelian physiognomy and the alienist searching out degeneracy in the criminal. All of these questions were raised within the context slavery in an age of Enlightenment.

3. Discourses on the stability of society, or social inertia (e.g., Parsonian sociology, the morality of community described by Nietzsche in the *Genealogy of Morals*, the rhetorics of stability, progress and degeneration which permeate the sociological discourses on the ends and uses of society). If one understands how these relations stand together, then we can begin to understand the relation between the scientific ideologies of race and the later sociological theories. As the definition of sociology narrowed over time there became a need to clean up its pantheon – Sumner, Spencer, Comte, Giddings, Cooley, Sorokin, Lombroso, etc. can all be listed amongst the missing idols. There absence does not mean that the relation between sociology and the government of populations has changed. In this same way, the fact that few remember the work of the American School or recognize the significance of Darwin's *Origin of Species* in countering polygenism does not mean that polygenic theories do not continue to haunt us today. We can discern in media representations of everyday life how this weighs like a nightmare on the mind of the living today.

Some might object to mentioning race in terms of discourse. The serious objection is that these discourses on nature, the forces of life and stability are not the same; that they have different histories, obey different laws, contain different errors, etc. This might be true. But these very different discourses, with such different histories, come together or intersect with each other in our time. In so far as this is correct, then these seemingly dispersed discourses meet in the process of social reproduction, in an apparatus here or an institution there. A further objection might be raised that the experience of racism is more important than the development of our classifications of race; that the content of the suffering has been left to the side. This point is well taken, but these discourses on human variety have their own origin in experience, in the experience of otherness that forced the European to define himself as man, of the encounter with someone different, and in the results produced by the belief in this supposed superiority. These differences are still often classified in accordance with what we believe to be the fundamental categories of *racial* difference and *scientific* truth. The experience of racialism was certainly taken as a fact within the realm of natural history. Indeed, it is a physical phenomenon, but the representation of that experience is, in essence and of necessity, social. It goes on between humans as they make their world, and it usually goes on behind their backs. It appears to be 'just the way the world is'. This is what gives the scientific ideology of race the aegis of truth and it is from this truth that scientific ideologies of race derive their 'authority'.

We should not be so arrogant as to think of polygenesis and Darwin's intervention as the triumph of reason over false science, for with the triumph came a shift from the antiquity of races to the antiquity of man and also new forms of knowledge such as degeneracy and eugenics, which wrapped themselves in the aegis of a Spencerian or Hegelian adaptation of Darwin's theory, all the while never attaining the scientific status of Darwinism. New forms of control rely on new systems of classification which have never quite left behind those of the late period of natural history. These are not new forms of unreason, and neither was polygenesis merely a false and wretched knowledge that was an ideological perversion of reason. It constituted scientific reason regarding man. Our present everyday knowledge of race owes much to it, but to the same degree so too do our own formations of biology and sociology rest upon it. This is why an inevitable conclusion to this book is the observation that to really understand the truth of race, the question 'What is race?' is irrelevant. The question that should be asked is 'What are ideologies that underlie the sciences of life and society?'

NOTES

Introduction

1. 'Scientific ideologies are explanatory systems that stray beyond their own borrowed norms of scientificity.' This is precisely why the history of science is really about the history of truth and error, and not falsity or false consciousness. 'Scientific ideology is not to be confused with false science, magic, or religion. Like them, it derives its impetus from an unconscious need for direct access to the totality of being, but it is a belief that squints at an already instituted science whose prestige it recognizes and whose style it seeks to imitate.' Which is to say that a scientific ideology will persist until such time as an adjacent discipline demonstrates its potential contribution to – and alignment with – an established disciplinary knowledge. G. Canguilhem, *Ideology and Rationality in the History of the Life Sciences* (Cambridge, MA: MIT Press, 1988), p. 38.
2. M. Serres, cited in ibid., p. 34.
3. Canguilhem, *Ideology and Rationality*, p. 39.
4. M. Horkheimer and T. Adorno, *Dialectic of Enlightenment: Philosophical Fragments* (1947; Stanford, CA: Stanford University Press, 2002), p. 8.
5. Ibid., p. 182.
6. A. Smedley, *Race in North America: Origin and Evolution of a Worldview* (Boulder, CO: Westview Press, 1993), p. 40.
7. 'The questions before us at this time are – 1. What is a species? 2. Are species permanent? 3. What is the basis of variations in species?' J. D. Dana, 'Thoughts on Species', *American Journal of Science and Arts*, 24 (1857), pp. 305–16, on p. 305.
8. Ideology is not merely the symbolic representation of social production; it is present in every moment in the process of production and accumulation, in every movement, thought, sound and gesture. Ideology is found in the discourses, technologies and moralities of everyday life produced by the social relations of capital. The mystification of social conflicts lies in the production and commodification of desire in everyday life, since part of capitalist social production is given over to the production of desire itself (see Horkheimer and Adorno, *Dialectic of Enlightenment*). Desire, especially the desire for one's own repression, is a social relation located 'in the particular social character of the labor that produces them' (K. Marx, *Capital*, 3 vols (New York: Penguin, 1976–81), vol. 1, p. 77.
9. C. Darwin, *The Descent of Man*, 2nd edn (1874; New York: Prometheus Books, 1998), p. 188.
10. G. de Santillana, *Reflections on Men and Ideas* (Cambridge, MA: MIT Press, 1968), pp. 82–3. For a notable example of such reductionism, see M. Heidegger's *Early Greek Thinking* (San Francisco, CA: Harper & Row, 1984).

11. Hesiod, *Works and Days*, 70–85. Hesiod's work is often characterized as either the rendition of a degenerative series or as an argument for cycles of history. In Hesiod's scheme, there could be no talk of ancient and modern within each epoch. We and the Greeks are contemporaries, all belonging to the same terrifying era, 'the time of iron' and for us, perhaps, more easily thought of as an era of terror.

12. Xenophon's *aphorme* is translated as 'capital' following Lyddell and Scott, 'starting-point, esp. in war, base of operations, 2. generally, starting-point, origin, occasion or pretext; 3. means with which one begins a thing, resources; 4. capital of a banker, etc.'. H. G. Liddell and R. Scott, *A Lexicon Abridged from Liddell and Scott's Greek–English Lexicon* (Oxford: Clarendon Press, 1958), p. 121.

13. Xenophon, *Ways and Means*, IV.21.

14. See A. Kojeve, *Introduction to the Reading of Hegel: Lectures on the Phenomenology of Spirit*, assembled R. Queneau; ed. A. Bloom; trans. J. H. Nichols, Jr (Ithaca, NY: Cornell University Press, 1980).

15. Aristotle, *The Politics of Aristotle*, ed. and trans. E. Barker (New York: Oxford University Press 1958), I.1254a. In *Prophesying by Dreams*, we find Aristotle refuting the notion that dreams are sent by the Gods because 'the power foreseeing the future and of having vivid dreams is found in persons of the inferior type, which implies that God does not send their dreams' (II.15). Aristotle goes on to say, quite rightly, that 'the principle ... expressed in the gamblers maxim: if you make many throws your luck must change' holds true too for dream interpreters as well. See also T. Lockwood, 'Is Natural Slavery Beneficial?', *Journal of the History of Philosophy*, 45:2 (2007), pp. 207–21.

16. Aristotle, *The Politics*, I.1254b.

17. See S. J. Gould's *Ontogeny and Phylogeny* (Cambridge, MA: Belknap Press of Harvard University Press, 1977) on the concept of recapitulation and its wide deployment in scientific discourse in the nineteenth century.

18. Aristotle's discourse on the bad mentality and deformed physique of the slave is absent from Xenophon's *Oeconomicus*. Instead, for Xenophon the fundamental differences between the master and the slave are of power and economy. The slave can be as moral and just – as 'good and beautiful' as any master. In the *Oeconomicus*, the master becomes a master because he is educated and trained to become a master. The master is not born a master. The master does not originate in any one people. The relation of master and slave is one of constant, and historical, change. The slave is no different either physically or mentally. Nor are men naturally superior to women, but both can be equally moral and just, or impious and immodest. The master performs his duties as a simple requirement of living an ethical life. 'In the *Oeconomicus*, there is no natural hierarchy among human beings according to gender, race or class ... Although Xenophon, like his contemporaries, took slavery for granted and assumed the system could be lucrative, he did not have a theory of natural slavery'. S. B. Pomeroy, *Xenophon Oeconomicus: A Social and Historical Commentary, with new English translation* (Oxford: Clarendon Press, 1994), p. 66.

19. G. W. F. Hegel, *The Philosophy of History* (1830–1), ed. C. J. Friedrich (New York: Dover Publications, 1956), pp. 96–100.

20. G. W. F. Hegel, *Hegel's Philosophy of Right* (1821), ed. T. M. Knox (New York: Oxford University Press, 1952), p. 261.

21. E. Gibbon, *The History of the Decline and Fall of the Roman Empire* (1783), ed. D. M. Milmam, M. Guizot and W. Smith (New York: Nottingham Society, 1845), pp. 256–7, citing Tacitus, *Annals*, XI.24. While the nineteenth-century citizen might well believe that the fall of Rome was caused merely by decadence and degeneracy, they certainly

looked to Tacitus for a description of the German, but also for his denouncing of the moral corruption of an imperial era. Less noticed by these same readers was Tacitus's own view, set forth on the first page of his history, that the moral corruption began with the foundation of imperial rule, and in the prohibition to expand the geographical range of Roman rule, as was dictated by Augustus in his will. The corruption was less about who was Roman, but rather, how are Romans to rule and be ruled. The decline originated in tyranny both in terms of social revolution and in the last testament of a tyrant. Tacitus admired the historians of the Republican era because they wrote of the people, unlike those of his day who produced mere biographies of the rulers. History had become, to him, not a means to transmit knowledge to the future but a discipline alternately servile and hostile to the state.

22. T. Frank, 'Race Mixture in the Roman Empire', *American History Review*, 21 (1916), pp. 689–708, on p. 705. Pritim Sorokin in his *Social and Cultural Mobility* (New York: Free Press, 1959), which founded the sociological study of social stratification, approvingly quotes this passage from Frank and follows it with citations to similar passages in le Bon and Gorbineau (pp. 328–30).

23. 'Notwithstanding the propensity of mankind to exalt the past and to depreciate the present, the tranquil and prosperous state of the empire was warmly felt, and honestly confessed, by the provincials as well as Romans. They acknowledged that the true principles of social life, laws, agriculture, and science, which had been first invented by the wisdom of Athens, were now firmly established by the power of Rome, under whose auspicious influence the fiercest barbarians were united by an equal government and common language. They affirm, that with the improvement of arts, the human species was visibly multiplied. They celebrate the increasing splendour of the cities, the beautiful face of the country, cultivated and adorned like an immense garden; and the long festival of peace, which was enjoyed by so many nations, forgetful of their ancient animosities, and delivered from the apprehension of future danger.' Pliny, cited in Gibbon, *The History of the Decline and Fall of the Roman Empire*, vol. 1, p 285, n. 109, see also Pliny *Natural History*, III.5.37–40.

24. Writing about the same time as Pliny (about 69 BC), Tacitus refrains from mentioning any monstrous races of men: 'The rest of what I have been able to collect is too much involved in fable, of a color with the accounts of the Hellusians and the Oxionians, of whom we are told, that they have a human face, with the limbs and bodies of wild beasts. But reports of these kind, unsupported by proof, I shall leave to the pen of others.' Tacitus, *Germania*, 46; see *A Treatise on the Situation, Manners, and People of Germany*, in *The Works of Tacitus*, ed. A. Murphy, 6 vols (Philadelphia, PA: Edward Earle, 1813), vol. 5, p. 259.

25. J. B. Friedman, *The Monstrous Races in Medieval Art and Thought* (Cambridge, MA: Harvard University Press, 1981), p. 8.

26. Pliny, *Natural History*, ed. H. Rackham, 10 vols (Cambridge, MA: Loeb Classical Library, 1942–52), vol. 2, pp. 510–13. Hodgen paraphrases the beginning of the quotation in order to bolster the contention that Pliny was simply uninterested in 'cultural phenomena' and his statements on culture amounted to mere epithets. This is how Hodgen gives the passage: 'At the outset, the numerical dimensions of the task of describing the varieties of human activity were startling and discouraging. Human cultures, [Pliny] remarked, were beyond counting and almost as numerous as the groups of mankind. As a result, there were hundreds of groups, tribes, and peoples, whose names are mentioned and who are located geographically with some care, but nothing more.' M.

T. Hodgen, *Early Anthropology in the Sixteenth and Seventeenth Centuries* (Philadelphia, PA: University of Pennsylvania Press, 1964), p. 36.

27. For example, Pliny tells the story of the twin slaves bought by Antony: 'the slave dealer Toranius sold to Antony after he had become one of the triumvirate two exceptionally handsome boys, who were so identically alike that he passed them off as twins, although one was a native of Asia and the other of a district North of the Alps. Later the boy's speech disclosed the fraud, and a protest was made to the dealer by the wrathful Antony, who complained especially about the large amount of the price (he had bought them for 200,000 sesterces); but the crafty dealer replied that the thing protested about was precisely the cause of his having charged so much, because there was nothing remarkable in a likeness between any pair of twin brothers, but to find natives of different races so precisely alike in appearance was something priceless; and this produced in Antony so great a feeling of good fortune that he, the inflictor of great atrocities, who had just before been hurling threats and abuse, suddenly found no other property better displayed his high station and good fortune.' Pliny, *Natural History*, ed. Rackham, vol. 2, pp. 543–5.

28. 'Openness' is a very relative term in this circumstance and the restrictions on being a Roman were based on relatively different criteria than the openness and closeness of our own time.

29. Seneca *Epistulae*, 47. In the *Saturnalia* of Macrobius, the discussion between Evangelus and Praetextatus goes along similar lines. Evangelus – thought by some commentators to be a have been a Christian– is appalled by the earlier suggestion that religion allows slaves 'taking their meals with their masters ... as if the gods would take any account of slaves or as if any sensible man would disgrace his house by keeping such low company in it' (*Saturnalia*, I.2). Praetextatus, a senator, replies that Evangelus refuses 'to reckon the slave a human being' and that he is in error, just as is his view that the gods take no notice of the condition of slaves. His brief discourse treats the humanity of slaves and their qualities as no less than those of their masters, nor did gender affect the basic humanity of the person. The discussion is very much in the spirit of Seneca, who Macrobius paraphrases, and Pliny. Macrobius, *The Saturnalia*, ed. P. V. Davies (New York: Columbia University Press, 1969), pp. 74–83.

30. Foucault makes much the same argument regarding the importance of language in natural history: 'When Jonston wrote his *Natural History of Quadrupeds*, did he know anything more about them than Aldrovandi did, a half-century earlier? Not a great deal more, the historians assure us. But that is not the question. Or, if we pose it in these terms, then we must reply that Jonston knew a great deal less than Aldrovandi. The latter, in the case of each animal he examined, offered the reader, and on the same level, a description of its anatomy and the methods of capturing it; its allegorical uses and mode of generation; its habitat and legendary mansions; its food and the best ways of cooking its flesh. Jonston subdivides his chapter on the horse under twelve headings: name: anatomical pars, habitat, ages, generation, voice, movement, sympathy and antipathy, uses, medicinal uses. None of this was omitted by Androvandi, and he gives a great deal more besides. The essential difference lies in what is missing in Jonston. The whole of animal semantics has disappeared, like a dead and useless limb. The words that had been interwoven in the very being of the beast have been unraveled and removed: and the living being, in its anatomy, its form, its habits, its birth and death, appears as though stripped naked. Natural History finds its focus in the gap that is now opened up between things and words.' M. Foucault, *The Order of Things: A History of the Human Sciences* (New York: Vintage Books, 1970), pp. 129–30.

31. 'The labours of Gregor Mendel have been a long time coming into their own. After prolonged experiments he wrote a paper, giving a clear and concise account of his results,
and sent it to the able botanist Nageli. This met with no response. Undaunted at first, he
sent it to a local scientific journal in 1865, where it languished in obscurity. Meeting with
no recognition, he died a disappointed and embittered man. Despite all appearances to
the contrary, he frequently remarked, "Meine Zeit wird schon kommen." The time came,
but it came to him as it came to Semmelweis, too late, for his body lay in the grave ... The
shores of the world of science are strewn with the wrecks of prophecies that remained
unheeded because many who might have recognized their value were either preoccupied
with conceptions with which they clashed or were thick in the fray with other scientists.
How many, except de Candolle, cared for Goethe's divination of the nature of vegetable
organism or for his anticipations of colour-theory? How many cared for the prophecy of
Marcus Antoninus Plenicz of the germ theory of disease, anticipating Pasteur by almost
a hundred years? How many cared when in 1846 Rasori announced that parasites produced fevers ... How many to the year 1900 cared for the outcome of Gregor Mendel's
experiments?' R. H. Murray, *Science and Scientists in the Nineteenth Century* (1925; New
York: Macmillan Co., 1988), p. 326.
32. O. Neuberger, *Astronomy and History: Selected Essays* (New York: Springer-Verlag,
1983), p. 3.
33. Marx, *Capital*, vol. 1, pp. 439–54.
34. F. Nietzsche, *On The Genealogy of Morals; Ecce Homo (1887–1888)*, ed. W. Kaufmann
and R. J. Hollingdale (New York: Vintage Books, 1969), p. 86.
35. B. R. Brown, 'City without Walls: Notes on Terror and Terrorism', *Situations: Project of
the Radical Imagination*, 2:1 (2007), pp. 53–82.

1 Classification and the Species Question

1. H. Ritvo, 'Zoological Nomenclature and the Empire of Victorian Science', in B. Lightman (ed.), *Victorian Science in Context* (Chicago, IL: University of Chicago Press, 1997),
pp. 334–53, on p. 335. See also H. Ritvo, *The Animal Estates: The English and Other
Creatures in the Victorian Age* (Cambridge, MA: Harvard University Press, 1987).
2. Ritvo, 'Zoological Nomenclature', p. 335.
3. L. Koerner, *Linnaeus: Nature and Nation* (Cambridge, MA: Harvard University Press,
1999).
4. See S. Freud, *Civilization and its Discontents*, ed. J. Strachey (New York: W. W. Norton & Co., 1961). The original German *Kultur* is invested with the double meaning of
culture and civilization, as Strachey notes: 'The original title chosen by Freud was *Das
Ungluck in the kultur* ("*Unhappiness in Civilization*"); but Ungluck was later altered to
"Unbehagen" – a word for which it is difficult to choose an English equivalent, though
the French "malaise" might have served. Freud suggested "*Man's Discomfort in Civilization*" in a letter to his translator, Mrs. Riviere; but it was she herself who found the ideal
solution in the difficulty in the title that was finally adopted' (p. 6). Freud himself says
in the *Future of an Illusion*: 'Human civilization, by which I mean all those respects in
which human life has raised itself above it animal status and differs from the life of the
beasts – and I scorn to distinguish between culture and civilization – presents as we
know, two aspects to the observer. It includes on the one hand all the knowledge and
capacity that men have acquired in order to control the forces of nature and extract its
wealth for the satisfaction of human needs, and on the other to adjust the relations of

men to one another and especially the distribution of the available wealth.' S. Freud, *Future of an Illusion*, ed. J. Strachey (New York: W. W. Norton & Co., 1961), p. 6.

5. A. Gschwendtner (dir.), *Der Menschen Forscher [The Anthropologist]* (1992). See also M. Wood, *Hitler's Search for the Holy Grail* (Mayavision International, 1998); G. Mosse, *Nationalism and Sexuality: Respectability and Abnormal Sexuality in Modern Europe* (New York: Howard Fertig, 1985); W. L. Shirer, *The Rise and Fall of the Third Reich: A History of Nazi Germany* (New York: Fawcett Crest, 1950).

6. See J. Ellis, *The Social History of the Machine Gun* (London: Cressett Press, 1971).

7. M. Omi and H. Winant, *Racial Formation in the United States: From the 1960s to the 1990s*, 2nd edn (New York: Routledge, 1994).

8. V. Rydberg, *Teutonic Mythology* (London: S. Sonnenschein & Co., 1889), p. 8. Max Muller, with whom the term originated, came to disown its use as having any relation to racial groupings. See M. Muller, *Chips from a German Workshop*, 3 vols (New York: Charles Scribner & Co., 1872).

9. Rydberg, *Teutonic Mythology*, pp. 17, 19.

10. J. P. Mallory, 'A History of the Indo-European Problem', *Journal of Indo-European Studies*, 1:1 (1973), pp. 21–65, on p. 22.

11. See note 5 above. See also C. Renfrew, *Archaeology and Language: The Puzzle of Indo-European Origins* (New York: Cambridge University Press, 1987); and H. A. Pringles, *The Master Plan: Himmler's Scholars and the Holocaust* (New York: Hyperion, 2006).

12. J. Parsons, *Remains of Japhet: Being Historical Enquiries into the Affinity and Origin of the European Languages* (London: for the author, 1767), p. 260.

13. W. Jones, *Eleven Discourses* (London: Thubner & Co., 1875), p. 18.

14. Foucault, *The Order of Things*, pp. 340–3.

15. J. Grimm, cited in G. P. Gooch, *History and Historians in the Nineteenth Century* (London: Longmans, Green, & Co., 1920), pp. 54–5.

16. C. Lyell, *Sir Charles Lyell's Scientific Journals on the Species Question* (1860), ed. L. G. Wilson (New Haven, CT: Yale University Press, 1970), pp. 410–11.

17. Latham, cited in Anon., 'The Aryan Question as it Stands Today: May Not the Original Home of the Indo-Germanic Peoples have been in Europe?', *New Englander and Yale Review*, 15:252 (1891), pp. 206–35, on p. 213.

18. Mallory, 'A History of the Indo-European Problem', p. 29.

19. 'Privileged over all others by the beauty of their blood and by the gifts of intelligence.' F. J. Pictet, *Les Origines indo-européennes, ou les aryas primitifs, essai de paleontologie linguistiqoe*, 2 vols (Paris: J. Cherbuliez, 1859–63), vol. 1, p. 7.

20. Morris, quoted in Mallory, 'A History of the Indo-European Problem', p. 33. Nor was this view obviously confined to the nineteenth century. Toynbee was at his most Hegelian when he wrote that 'The Black races alone have not contributed positively to any civilization – as yet'. A. Toynbee, *A Study of History* (New York: Dell Publishing, 1969), p. 74. Compare Hegel's statement: 'Africa proper, as far as History goes back, has remained – for all purposes of connection with the rest of the world – shut up; it is the Gold-land compressed within itself – the land of childhood, which lying beyond the day of self-conscious history, is enveloped in the dark mantle of night ... it has no historical part of the World; it has no movement or development to exhibit'. *The Philosophy of History*, pp. 91–9.

21. '[L]anguage in the 19th century, throughout its development and even in its complex forms, was to have an irreducible expressive value, for if language expresses, it does so not in so far as it is an imitation and duplication of things, but in so far as it manifests and

translates the fundamental will of those who speak it ... language is no longer linked to civilizations by the level of learning to which they have attained ... but by the mind of the peoples who have given rise to it, animated it, and are recognizable in it. Just as a living organism manifests, by its inner coherence, the functions that keep it alive, so language, in the whole architecture of its grammar, makes visible the fundamental will that keeps a whole people alive and gives it the power to speak a language belonging solely to itself ... In any language, the speaker, who never ceases to speak in a murmur that is not heard although it provides all the vividness of the language, is the people. Grimm thought that he overheard such a murmur when he listened to the *altdeutsche Meistergesang*, and Raynouard when he transcribed the *Poesies originales des troubadours*. Language is no longer linked to the knowing of things, but to men's freedom: "Language is human: it owes its origin and progress to our full freedom; it is our history, our heritage". By defining the internal laws of grammar, one is simultaneously linking language and the free destiny of men in a profound kinship. Throughout the 19th century, philology was to have profound political reverberations.' Foucault, *The Order of Things*, pp. 290–1.

22. Renfrew, *Archaeology and Language*, pp. 14–15.
23. Gooch, *History and Historians*, p. 59. Cf. 'Genuinely poetic projection is the opening up or disclosure of that into which human being as historical is cast. This is the earth and, for a historical people, its earth, the self-closing ground on which it rests together with everything that it already is, though still hidden from itself. It is, however, its world, which prevails in virtue of the relation of human being to the unconcealedness of Being. For this reason, everything with which man is endowed must, in the projection, be drawn up from the closed ground and expressly set upon this ground as the bearing ground. All creation, because it is such a drawing-up, is a drawing as of water from a spring ... History is the transporting of a people into its appointed task as entrance into that people's endowment ... [the origin of the work of art] is history in the essential sense that it grounds history ... the origin of the work of art is the origin of both the creators and the preservers, which is to say of a people's historical existence.' M. Heidegger, 'The Origin of the Work of Art' (1935), in A. Hofstadter (ed.), *Poetry, Language, Thought* (New York: Harper & Row, 1971), pp. 75–8.
24. Gooch, *History and Historians*, p. 55.
25. J. Grimm, *Teutonic Mythology* (1844), ed. J. S. Stallybrass (New York: Dover Publications, 1966), p. 93.
26. Rydberg, *Teutonic Mythology*, p. 11.
27. Grimm, cited in Gooch, *History and Historians*, pp. 54–5. Padraic Colum and Joseph Campbell were drawn to just this aspect of the Grimm's work: 'We have another past besides the past that history tells us about, a past which is in us, in individuals, more livingly than the recorded past ... It is this long past ... that comes to us in these and other traditional stories. With it certain things are restored to our imagination. Wilhelm Grimm, who knew much more about the inwardness of these stories than the philologists and the historians of culture who were to comment on them, was aware of "fragments of belief dating back to most ancient times, in which spiritual things are expressed in a figurative manner." "The mythic element," he told us, "resembles small pieces of a shattered jewel which are lying strewn on the ground all overgrown with grass and flowers, and can only be discovered by the most far-seeing eye." "Their signification has long been lost, but it is still felt." he says, "and imparts value to the story".' P. Colum, 'Introduction', in *The Complete Grimm's Fairy Tales* (New York: Partheon Fairy Tale and

Folklore Library, 1972), pp. vii–xiv, on p. xiv. For an elaboration of this view, see also J. Campbell, 'Folkloristic Commentary' (1944), in the same volume, pp. 833–64.

28. E. Prokosch, *A Comparative Germanic Grammar*, William Dwight Whitney Linguistic Series (Philadelphia, PA: Linguistic Society of America and University of Pennsylvania, 1939), pp. 55–7.

29. Ibid., p. 57.

30. Rydberg, *Teutonic Mythology*, pp. 14, 18.

31. Ibid., p. 18, 19, 21.

32. T. H. Markey, in R. K. Rask, *A Grammar of the Icelandic or Old Norse Tongue* (1811), ed. T. H. Markey (Amsterdam: Benjamins, 1976), pp. xvii, xi, x. Rask also recognized that the Old Icelandic sagas described actual Viking expeditions to Labrador, Newfoundland and Nova Scotia.

33. '[T]he body of the Earth is only the shape of this organism ... The members of this organism do not contain, therefore, the generality of the process within themselves, they are the Particular individuals, and constitute a system whose forms manifest themselves as members of the unfolding of an idea, whose process of development is a past one.' G. F. W. Hegel, *Philosophy of Nature*, ¶¶261–2, at http://www.marxists.org/reference/archive/hegel/ [accessed 1 May 2010].

34. Gould, *Ontogeny and Phylogeny*, pp. 18, 100–9.

35. Rask, *A Grammar of the Icelandic or Old Norse Tongue*, p. 227.

36. Grimm, *Teutonic Mythology*, p. ix.

37. 'The position taken by most German archaeologists regarding the Aryan question in the 1930's is well known ... 1936 produced a bumper crop of German articles on the IE problem, all of whom sought the homeland in Northern Europe'. Mallory 'A History of the Indo-European Problem', p. 48.

38. Saint-Hilaire, cited in J. Bachman, 'Unity of the Human Race', *Southern Quarterly Review*, 9:17 (January 1846), pp. 1–57, on p. 36.

39. F. de Saussure, *Course in General Linguistics* (Chicago, IL: Open Court, 1983), p. 221. See also Prokosch's introductory essay, 'The External History of the Germanic Languages', in his *A Comparative Germanic Grammar*, pp. 21–34.

40. Hegel, *The Philosophy of History*, pp. 98–9.

41. Ibid., p. 92.

42. Ibid., pp. 91–2. Cf. 'European annexation waited upon exploration. Africa was the "Dark Continent," and until the darkness was lifted it was not converted. About the middle of the century the darkness began to disappear. Explorers penetrated farther and farther into the interior, traversing the continent in various directions, opening a chapter of geographical discovery of absorbing interest ... by 1880, the scientific enthusiasm and curiosity, the missionary and philanthropic zeal of Europeans, the hatred of slave-hunters who plied their trade in the interior, had solved the great mystery of Africa, the map showed rivers and lakes where previously where previously all had been blank. Upon discovery quickly followed appropriation.' C. D. Hazen, *Modern European History* (New York: Henry Holt & Co., 1917), pp. 308–9.

43. Hegel, *The Philosophy of History*, p. 96. We might note that 300 years before Hegel both al-Idrisi and Ibn Khaldun described large parts of the interior of Africa, at least above the equator. They asserted that the lack of large civilizations was caused by climate.

44. Ibid., p. 93.

45. See S. Buck-Morss, 'Hegel and Haiti', *Critical Inquiry*, 26 (2000), pp. 821–67.

46. Hegel, *The Philosophy of History*, p. 95.

47. Ibid., pp. 96–7. Compare the final passage with the opening of J. G. Frazer's *Golden Bough: A Study in Comparative Religion*, 2 vols (London: Macmillan, 1890) and his analysis of the myth of the King of the Wood.

48. Hegel, *The Philosophy of History*, p. 96.

49. Ibid., p. 98.

50. Ibid., p. 98.

51. Ibid., p. 96. It was often said that we will know if the Negro is fully human by his ability to fight, and so the degree of civilization is to be measured by the discipline of the army. It is an idea which appears often in American history, and it seems that in every war until the army was integrated, black regiments had to prove the point in some of the bloodiest combat.

52. Ibid., p. 99.

53. See S. Buck-Morss, *Hegel, Haiti, and Universal History* (Pittsburgh, PA: University of Pittsburgh Press, 2009) for one of the most insightful readings. Buck-Morss has as her focus the early period of Hegel's work, particularly the Jena period which coincided with the Haitian Revolution and the Napoleonic War.

54. Hegel, *The Philosophy of History*, p. 99.

55. Cf. Foucault, *The Order of Things*, pp. 125–65.

56. Never mind that for the Greeks attire progressed in the opposite direction from the wearing of garments and weapons to the wearing of loose-fitting robes and public nudity in the polis. See, for example, Thucydides's *Peloponnesian War*, I.6.3.

57. Foucault, *The Order of Things*, pp. 344–87.

58. Hoppe, cited in J. S. Slotkin (ed.), *Readings in Early Anthropology: A Comprehensive Anthology of Pre-Scientific Writings on the Nature, Origin, History and Behavior of Man* (Chicago, IL: Aldine Publishing Co., 1965), p. 181.

59. Koerner, *Linnaeus*, p. 283. Koerner observes that writing a dissertation under Linnaeus was a singular experience, where the student literally sat at the feet of the master as his dissertation was dictated to him.

60. Soderberg, cited in Slotkin (ed.), *Readings in Early Anthropology*, p. 183.

61. Linné, cited in ibid., pp. 180, 179, 281.

62. Ibid., pp. 179–80.

63. A. O. Lovejoy, *The Great Chain of Being: A Study of the History of an Idea* (Cambridge, MA: Harvard University Press, 1936), p. 235. One such traveller, Sir John Ovington in his *Voyage to Suratt* (London: Jacob Tonson, 1696), stated that the '*Hotantots* ... are the very Reverse of Human kind ... so that if there's any medium between a Rational Animal and a beast, the *Hotantot* lays the fairest Claim to that Species' (p. 489). See also J. Leidy, 'On the Hair of a Hottentot Boy', *Journal and Proceedings of the Academy of Natural History of Philadelphia* (1847), p. 7; S. G. Morton, 'Some Observations on the Bushman Hottentot Boy', *Journal and Proceedings of the Academy of Natural History of Philadelphia* (1848), p. 5; N. Hudson, 'Hottentots and the Evolution of European Racism', *Journal of European Studies*, 34:4 (2004), pp. 308–32.

64. Koerner, *Linnaeus*, pp. 187–93.

65. G. L. Leclerc, comte de Buffon, *Natural History*, 10 vols (London: J. S. Barr, 1792), vol. 10, p. 350.

66. S. J. Gould, 'The Man Who Invented Natural History', Review of *Buffon* by Jacques Roger, trans. Sarah Lucille Bonnefoi, Cornell University Press, *New York Review of Books*, 45:16 (22 October 1998), pp. 83–6, on p. 86.

67. Buffon, cited in W. Lawrence, *Lectures on Physiology, Zoology, and the Natural History of Man* (London: James Smith, 1823), pp. 444–5.
68. Buffon, *Natural History*, vol. 5, p. 1.
69. T. S. Savage and J. Wyman, 'Notice of the External Characteristics and Habits of Troglodytes Gorilla, a New Species of Orang from the Gaboon River', in J. C. Burnham (ed.), *Science in America: Historical Selections* (New York: Holt, Rinehart & Winston, 1971), pp. 115–27, on pp. 126–7. The medieval explorer Hanno had reported that on an island within an island, he had found a 'people ... whose bodies were hairy and whom our interpreters called *Gorillae*'. Savage and Wyman note that Hanno referred to these as men and women whom he and his companions pursued, capturing 'three women, but they attacked their conductors with teeth and hands and could not be prevailed upon to accompany us. Having killed them, we flayed them and brought their skins with us to Carthage' (p. 117). See *The Voyage of Hanno Translated, and Accompanied with the Greek Text*, trans. T. Falconer (London: T. Cadell, Jr & Davies, 1797), p. 13.
70. Buffon, *Natural History*, vol. 5, p. 167.
71. 'In the New World, he famously insisted, much to the irritation of Thomas Jefferson, an inferior climate inevitably produced a degenerate anthropology, a mediocre zoology, and a substandard botany. But these were not inherent; they were derived. So, while on the surface it might look as though "the Negro, the Laplander, and the White were really different species," in fact "those marks which distinguish men who inhabit different regions of the earth, are not original, but purely superficial. It is the same identical being who is varnished with black under the Torrid Zone, and tawned and contracted by extreme cold under the Polar Circle."' D. Livingstone, *Adam's Ancestors: Race, Religion, and the Politics of Human Origins* (Baltimore, MD: Johns Hopkins University Press, 2008), p. 56.
72. T. Jefferson to Dr John Manners, Monticello, 22 February 1814, 'Classification in Natural History', in *Letters*, ed. M. D. Peterson (New York: Literary Classics of the United States, 1984), pp. 1329–33, on pp. 1329–30.
73. T. Jefferson to Chastellux, 7 June 1785, 'On American Degeneracy', in ibid., pp. 799–803.
74. S. J. Gould, 'Ladders and Cones: Constraining Evolution by Canonical Icons', in R. B. Silvers (ed.), *Hidden Histories of Science* (New York: New York Review Books, 1995), pp. 37–68. Gould's work on the iconography of evolution shows that in many ways Buffon's expansive interpretation is much closer to Darwin's tangled one. Lacking an adequate mechanism to explain heredity, the genealogical tree was presented in the shape of a tree. Haeckel's illustration is particularly noteworthy in this case, but Rydberg's tree also comes easily to mind.
75. Jefferson to Chastellux, 'On American Degeneracy', pp. 800, 801, 803
76. T. Jefferson, *Notes on the State of Virginia* (1787), ed. W. Peden (New York: W. W. Norton & Co., 1954), p. 93.
77. Ibid., p. 101.
78. Ibid., pp. 70–1.
79. See Voltaire, 'Relation touchant un maure blanc', cited in Koerner, *Linnaeus*, p. 87, n. 37, as 'Relating to a White Moorish Amene of Africa in Paris in 1744'. Karl Marx's curious early verse 'The Viennese Ape Theatre in Berlin' (1837), in *Karl Marx, Frederick Engels: Collected Works*, 50 vols (London: Lawrence & Wishart; New York: International Publishers; Moscow: Progress Publishers, 1975–2004), vol. 1, p. 539. See also F. Spenser, 'Two Unpublished Essays on the Anthropology of North America by Benjamin Smith Barton', *Isis*, 68:4 (1977), pp. 567–73.

80. T. Jefferson, *The Jefferson Bible: The Life and Morals of Jesus of Nazareth extracted Textually from the Gospels* (St Louis, MO: N. D. Thompson Publishing Co., 1902), p. 50.
81. S. F. Haven, *Archaeology of the United States; or, Sketches, Historical and Bibliographical, of the Progress of Information and Opinion respecting Vestiges of Antiquity in the United States* (Washington, DC: Smithsonian Institution and G. P. Putnam & Co., 1856), p. 94.
82. Jefferson, *Notes on the State of Virginia*, p. 140.
83. Slotkin (ed.), *Readings in Early Anthropology*, pp. 189–91.
84. Savage and Wyman, 'Notice of the External Characteristics and Habits of Troglodytes Gorilla'.
85. In 1913, Webster gave the current view in this way: '*Races of Man:* At the dawn of history, the various regions of the world were already in the possession of many different peoples. Such characteristics as the shape of the skull, the features of the face, stature, or complexion, may serve to distinguish one people from another. Other grounds for distinction are found in language, customs, beliefs, and general intelligence. *Classifications of Race:* If we consider physical differences only, it is possible to classify the world's inhabitants into a few large groups or races. Each of these groups occupies, roughly speaking, its separate area of the globe. The most familiar classification is that which recognizes the Black or Negro race dwelling in Africa, the Yellow or Mongolian race whose home is in central and eastern Asia, and the White or Caucasian race of western Asia and Europe. Sometimes two additional divisions are made by including, as the Red race, the American Indians, and as the Brown race, the natives of the Pacific islands. *The White Race:* These separate racial groups have made very unequal progress in culture. The peoples belonging to the Black, Red, and Brown races are still either savages or barbarians as were the men of prehistoric times. The Chinese and the Japanese are the only representatives of the Yellow race that have been able to form civilized states. In the present, as in the past, it is chiefly the members of the White race who are developing civilization and making history.' H. Webster, *Ancient History* (New York: D. C. Heath & Co., 1913), pp. 23–5.
86. J. F. Blumenbach, 'On the Natural Variety of Mankind', in *The Anthropological Treatises of Johann Friedrich Blumenbach*, trans. and ed. T. Bendyshe (Boston, MA: Longwood Press, 1978), pp. 65–144, on pp. 98–9.
87. Ibid., pp. 98–9, 100.
88. Blumenbach, cited in T. Gossett, *Race: the History of an Idea in America* (Dallas, TX: Southern Methodist University, 1963), p. 39.
89. Herder, cited in A. Montagu, *Man's Most Dangerous Myth: The Fallacy of Race*, 5th edn, revised and enlarged (New York: Oxford University Press, 1974), p. 26.
90. Blumenbach, 'On the Natural Variety of Mankind', p. 99.
91. It is another irony that species do not all evidence increasing complexity, although the classifications of the taxonomists certainly display this tendency.
92. J. F. Blumenbach, *On the Natural Varieties of Mankind. De generis humani varietate nativa* (1776), ed. T. Bendyshe (New York: Bergman Publishers 1969), pp. 109–10.
93. Gould notes that the genealogy of the term 'degenerate' can be traced to Blumenbach's attempt to account for the diversity of humans by recourse to the argument that variety is the result of deviation from the original stock of humans, a degeneration ('de' from genus stock or line of descent) 'not intending, by this word, the modernist sense of deterioration, but the literal meaning of departure from the initial form of humanity at the time of the Creation'. S. J. Gould *The Mismeasure of Man*, 2nd edn (New York: W. W. Norton, 1999), p. 407.
94. Ibid., p. 410.

95. Ibid., p. 405.
96. Ibid., p. 410.
97. Blumenbach, cited in Bachman, 'Unity of the Human Race', p. 4.
98. Livingstone's *Adam's Ancestors* takes up much of the story of polygenism and religion in the nineteenth century. Our lines of analysis intersect quite nicely and are mutually supportive.
99. G. Cuvier, *A Discourse on the Revolutions of the Surface of the Globe, and the Changes Thereby Produced in the Animal Kingdom* (Philadelphia, PA: Carey & Lea, 1831), p. 11.
100. If there was already the ground for Darwin and the other great nineteenth-century thinkers, it can be found in the contributions of Buffon's geographical adaptation and genealogy; Cuvier's extinction, catastrophe, discontinuity and viewing the fossil record as history; Lyell's conceptions of geological change and the history of the earth; and Charles's own grandfather Erasmus Darwin's suggestions on evolution and change.
101. Cf. G. Stocking, *Race, Culture, and Evolution: Essays in the History of Anthropology: With a New Preface* (Chicago, IL: University of Chicago Press, 1982), p. 39.
102. Cuvier, *A Discourse on the Revolutions of the Surface of the Globe*, pp. 179–80.
103. C. Singer, *A History of Biology: A General Introduction to the Study of Living Things* (New York: Henry Schumann, 1951), pp. 279–80.
104. W. Coleman, *Georges Cuvier, Zoologist: A Study in the History of Evolution Theory* (Cambridge, MA: Harvard University Press, 1964), pp. 233–4.
105. Cuvier, *A Discourse on the Revolutions of the Surface of the Globe*, p. 220
106. 'Relics of bygone instruments of labor possess the same importance for the investigation of extinct economical forms of society as do fossil bones for the determination of extinct species.' Marx, *Capital*, vol. 1, p. 285.
107. J. L. Larson, *Reason and Experience Reason and Experience, the Representation of Natural Order in the Work of Carl von Linné* (Berkeley, CA: University of California Press, 1971), p. 152.
108. Cuvier, *A Discourse on the Revolutions of the Surface of the Globe*, p. 136.
109. P. Lester, 'L'anthropologie', in M. Daumas (ed.), *Histoire de la science* (Paris: Gallimard, 1957), p. 1367.
110. M. J. S. Rudwick, *Georges Cuvier, Fossil Bones, and Geological Catastrophes* (Chicago, IL: University of Chicago Press, 1997), p. 261.
111. One noted anthropologist, Carleton S. Coon, speculated that there are as many as 143 human races. See C. S. Coon with E. E. Hunt, Jr, *The Living Races of Man* (New York: Alfred A. Knopf, 1966).
112. Rudwick, *Georges Cuvier*, pp. 260–1.
113. Stocking, *Race, Culture, and Evolution*, pp. 29–31.
114. E. Norduskiold, *The History of Biology: A Survey* (New York: Alfred A. Knopf, 1921), p. 343.
115. J. P. Lamarck, *Philosophie zoologique: ou Exposition des considérations relative à l'histoire naturelle des animaux* (Paris: Dentu, 1809), p. ii.
116. Norduskiold, *The History of Biology*, p. 343.
117. W. P. D. Wrightman, *The Growth of Scientific Ideas* (New Haven, CT: Yale University Press, 1951), p. 403.
118. Horkheimer and Adorno, *Dialectic of Enlightenment*, p. 91.
119. Lyell, *Scientific Journals on the Species Question*, pp. 191, 429.
120. Ibid., p. 57.

121. 'They were dying slowly – it was clear. They were not enemies, they were not criminals, they were nothing earthly now – nothing but the black shadows of disease and starvation living confused in the greenish gloom. Brought from all the recesses of the coast in all the legality of time contracts, lost in uncongenial surroundings, fed on unfamiliar food, they sickened, became inefficient, and were then allowed to crawl away and rest.' J. Conrad, *Heart of Darkness and the Secret Sharer*, ed. F. Walker (New York: Bantam Books, 1969), p. 26. See also *An Outpost of Progress*, in *Selected Tales from Conrad*, ed. N. Stewart (London: Faber Paperbacks, 1977), pp. 14–42, on pp. 16–18, as Kayerts and Carlier depart for the trading station of the title to the contempt of the 'Managing Director of the Great Civilizing Company (since we know that civilization follows trade)'.
122. G. Cuvier, 'Mémoire sur l'ibis des anciens Egyptiens', in *Recherches sur les ossements fossiles de quadrupèdes: discours préliminaire* (Paris: GF-Flammarion, 1998), pp. 222–46, on pp. 245, 223.
123. Ibid., p. 246.
124. Ibid., p. 245.
125. S. G. Morton, *Crania Americana* (Philadelphia, PA: J. Dobson, 1839), p. 31.

2 Polygenesis and the Types of Mankind

1. In 1663 Winthrop became the first American to be elected to the Royal Society.
2. J. E. Dekay, 'Anniversary Address on the Progress on the Natural Sciences in the United States: Delivered before the Lyceum of Natural History of New York' (1826), in Burnham (ed.), *Science in America*, pp. 75–89, on p. 76.
3. Gossett, *Race*, p. 58.
4. Anon., 'The Original Unity of the Human Race – Pickering, Bachman, Agassiz', *New England and Yale Review*, 8 (1850), pp. 542–85, on p. 542.
5. The By-Laws of the Anthropological Society of New York state that '[i]ts special Object is the study of the History, Conditions, and Relations of the Aboriginal Inhabitants of America, and the Phenomena resulting from the contact of the various races and families on the American Continent, before and since the Discovery. 3. The Physical Characteristics, Religious Conceptions, and Systems of Men; their Mythology and Traditions; their Social, Civil, and Political Organizations and Institutions; their Language, Literature, Arts, and Monuments; their Modes of Life and their Customs, are specifically within the objects of the Institute', in *Journal of the Anthropological Institute of New York*, 1:1 (1871), p. 6.
6. W. Stanton, *The Leopard's Spots: Scientific Attitudes toward Race in America 1815–59* (Chicago, IL: University of Chicago Press, 1960), p. 196.
7. The late senator from New York and noted sociologist Daniel Patrick Moynihan illustrated this in a 1994 Senate hearing on teen pregnancy and single-mothers: 'It has something to do [with] a changed condition in biological circumstances. I mean, if you were a biologist, you could find yourself talking about speciation here.' G. Hodgson, *The Gentleman from New York: Daniel Patrick Moynihan* (Boston, MA: Houghton Mifflin, 2000).
8. J. C. Nott, 'The Unity of the Human Race', *Southern Quarterly Review*, 9:17 (January 1846), pp. 1–57, on p. 1. The cover of the copy in the collection of the American Museum of Natural History is inscribed '2nd article in reply to the Rev. Curtis of South Carolina'.
9. Ibid., pp. 1, 2.

10. Ibid., p. 3.
11. Blumenbach, cited in ibid., p. 4.
12. Virey, cited in ibid., p. 5. According to the World Register of Marine Species, varieties of the bread crumb sponge have been misidentified 56 times as separate species. BBC World Service, 25–6 June 2008.
13. Prichard, cited in Nott, 'The Unity of the Human Race', p. 14.
14. Nott, 'The Unity of the Human Race', pp. 20–1.
15. Saint-Hilaire, cited in ibid., p. 36.
16. See Livingstone, *Adam's Ancestors*. Polygenism continues to provide many with a means to make sense of human variety with a theory that rationally reconciles Christian theology with supremacist ideology. The period when the non-theological polygenic theories ascended to scientific orthodoxy and their survival within scientific and social theory cannot be explained on the basis of a need to reconcile reason with theology, nor by irrationality because, as Livingstone points out, polygenism had a rational basis in both theological criticism and anthropological investigation. The repudiation of their scientific work was not complete. The work of the polygenists would still fall within the bounds of normal science. That they are forgotten and the theory is today amongst the wretched subjects does not diminish the profound influence they maintained, nor that it was scientific polygenism that Darwin hoped would disappear. Even as he dealt a great blow to the scientific vitality of polygenism, he understood well the limitation of his work as a corrective to theological belief, which, while it can and does appropriate rational tools, remains as always unscientific.
17. Jefferson, *Notes on the State of Virginia*, pp. 150–5.
18. Nott, 'The Unity of the Human Race', pp. 40–1.
19. Muller does just the opposite, and lets physical anthropology stand in where the texts are lacking, see his essays concerning natural history and mythology in *Chips from a German Workshop*, vol. 2, pp. 1–141, 195–205, 248–356.
20. Stanton, *The Leopard's Spots*, p. 119.
21. Because of the work of the Christian abolitionists, it would be easy to overestimate the degree to which religious instruction and belief really opposed slavery despite the biblical support for the monogenic origins of humans. For example, this passage: 'In 2247 B.C. the sons of men banded themselves together to build the tower of Babel on the plain of Shinar, just below the hills of Armenia, where the two great rivers Euphrates and Tigris make the flats rich and fertile. For their presumption, God confounded their speech, and the nations first were divided. Ham's children got all the best regions; Nimrod, the child of his son Cush, kept Babel, built the first city, and became the first king. Canaan's sons settled themselves in that goodliest of all lands which bore his name; and Mizraim's children obtained the rich and beautiful valley of the Nile, called Egypt. All these were keen clever people, builders of cities, cultivators of the land, weavers and embroiderers, earnest after comfort and riches, and utterly forgetting, or grievously corrupting, the worship of God. Others of the race seem to have wandered further south, where the heat of the sun blackened their skins; and their strong constitution, and dull, meek temperament, marked them out to all future generations as a prey to be treated like animals of burden, so as to bear to the utmost the curse of Canaan.' C. M. Yonge, *The Chosen People: A Compendium of Sacred and Church History for School-Children* (London: J. & C. Mozley, 1859), p. 3.
22. Nott was well known in Mobile, where he maintained a thriving medical practice, and during this period he lived comfortably treating patients and conducting research on

tropical diseases such as Yellow Fever. See J. C. Nott, 'Yellow Fever Contrasted with Billious Fever – Reason for Believing it a Disease of Sui Generis ... Probably Insect or Animalcular Origin', *New Orleans Medical and Surgical Journal*, 4 (1848), pp. 563–601.

23. Nott, 'The Unity of the Human Race', p. 45.

24. Stanton, *The Leopard's Spots*, pp. 65–6; Anon. ('Philanthropist'), 'Vital Statistics of Negroes and Mulattoes', *Boston Medical and Surgical Journal*, 17 (1842), pp. 168–70.

25. V. L. Parrington, *The Romantic Revolution in America, 1800–1869* (New York: Harcourt Brace Jovanovich, 1954), pp. 98–103, on p. 98. *DeBow's Review* published extracts as 'The Hireling and the Slave', *Debow's Review, Agricultural, Commercial, Industrial Progress and Resources*, 18:2 (February 1855), pp. 185–8; 19:2 (August 1855), pp. 208–18. The complete poem can be found in W. J. Grayson, *Selected Poems by William J. Grayson*, comp. Mrs W. H. Armstrong (New York, Washington: Neale Publishing Co., 1907), pp. 21–79. See also DeBow's essays 'The Hireling and the Slave (continued)', *Debow's Review, Agricultural, Commercial, Industrial Progress and Resources*, 18:4 (April 1855), pp. 459–62; and 'The Hireling and the Slave – Illustrations from Mr. Grayson's Poem', *Debow's Review, Agricultural, Commercial, Industrial Progress and Resources*, 21:3 (September 1856), pp. 248–56.

26. Stanton, *The Leopard's Spots*, p. 58. Stanton's book is really unsurpassed in its telling of the story, an engrossing read and a singular narrative. This is not to say that Stanton's narrative is definitive in all respects. Certainly it contains both errors of its time and errors that come from being the first work of its kind. L. D. Stephens's recent *Science, Race, and Religion in the American South: John Bachman and the Charleston Circle of Naturalists, 1815–1895* (Chapel Hill, NC: University of North Carolina, 2000) has detailed some of these deficiencies quite well. Unfortunately, Stephens's book takes the tone of one settling old scores and is somewhat defensive about Stanton's lack of regard for the Charleston naturalists, except for Bachman. Still, Stephens does the yeoman's work of putting the scientific work of the time into perspective. Many of the conclusions in this chapter regarding the importance of the American School essentially agree with Stephens. Notice of his work came late to the writing of this material and so serves more as confirmation than inspiration.

27. E. Jarvis, 'Statistics of Insanity in the United States', *Boston Medical and Surgical Journal*, 27 (1842), pp. 116–21, 281–2, on pp. 281–2.

28. Stanton notes that the primary difference between 1830 and 1840 was that in the interim, the fields of vital statistics and public hygiene appeared. *The Leopard's Spots*, p. 54. The categories mentioned are terms taken from the 1840 Census form.

29. M. Anderson, *The American Census: A Social History* (New Haven, CT: Yale University Press, 1990), pp. 29–31. Anderson is not alone in ignoring Jarvis's first conclusions, for her sources do too. In 1960 Stanton had already noted the same trend amongst his commentators and even in his eulogy and obituaries. *The Leopard's Spots*, p. 212.

30. Stanton, *The Leopard's Spots*, p. 58.

31. Jarvis, 'Statistics of Insanity in the United States', pp. 116–21.

32. Ibid., pp. 281–2.

33. Anon., 'Table of Lunacy in the United States', *Hunt's Merchant's Magazine*, 8 (May 1843), pp. 460–2; Anon., 'Reflections on the Census of 1840', *Southern Literary Messenger*, 9:6 (June 1843), pp. 340–52.

34. Jarvis, 'Statistics of Insanity in the United States', on pp. 281–2. At the time, the census was conducted by the Department of State.

35. Calhoun to Mr Pakenham, Department of State, Washington, 18 April 1844, in *The Works of John C. Calhoun*, ed. R. K. Crallé, 6 vols (New York: D. Appleton, 1883), vol. 6, p. 347.

36. J. C. Calhoun, 'On the Reception of Abolition Petitions, Delivered in the Senate, February 6th, 1837 [Slavery as a "Positive Good"]', in ibid., vol. 2, pp. 625–33, on p. 632. Grayson's epic poem *The Hireling and the Slave* was once such approving response to Calhoun's challenge.

37. Gliddon to Calhoun, quoted in Stanton, *The Leopard's Spots*, p. 213.

38. E. Jarvis, 'Increase of Human Life, III', *Atlantic Monthly*, 24:146 (1869), pp. 711–19, on p. 718.

39. Stanton, *The Leopard's Spots*, p. 228.

40. 'In 1830, Dr. Samuel G. Morton, of Philadelphia, already distinguished as a naturalist, having occasion to lecture on the form of the skull as exhibited in the five races of man, found himself unable to procure specimens of each for his illustrations. Impressed with this great scientific deficiency, he resolved upon making a collection himself. By the most indefatigable exertions, and a large expenditure of time and money, he succeeded in obtaining a cabinet of crania superior to any in the world. The cost to Dr. Morton was estimated at from ten to fifteen thousand dollars, and at the time of his death he had accumulated nearly a thousand human skulls, derived from all quarters of the globe.' Haven, *Archaeology of the United States*, p. 77; see also S. G. Morton, *Catalogue of Skulls of Man and the Inferior Animals in the Collection of Samuel George Morton* (Philadelphia, PA: F. Turner, 1844); and J. A. Meigs, *Catalogue of Human Crania, in the Collection of the Academy of Natural Sciences of Philadelphia, PA: Based upon the third edition of Dr. Morton's 'Catalogue of skulls'* (Philadelphia, PA: J. B. Lippincott & Co., 1857). Stanton, *The Leopard's Spots*, pp. 25–35, 37–46, 89–102, 139–45, 154–7, 170–92; Gould, *The Mismeasure of Man*, pp. 62–103.

41. 'The richness of the craniological collection you have so cheerfully brought together finds in you a worthy interpreter. Your works, Sir, are notable for the depth of the anatomical observations, by the detailed numeric reports and confirmations of organic ratios, by the absence of the poetic reveries which are the myths of the modern physiologic, and by the wide range of information that abounds in your "Introductory Essay." Coming as it does to me at an important moment in my own work, which will be published under the imprudent title *Cosmos*, I will benefit from the many insights on the distribution of the human races in your beautiful volume. Sacrifices to reach such great artistic perfection, and to produce a work that competes with it, one would search in vain throughout all of England and France.' A. von Humbolt to Morton, January 1844, in Haven, *Archaeology of the United States*, p. 78.

42. E. G. Squire, 'Announcement of Morton's Death to the Ethnological Society', *International Magazine of Literature, Art, and Science*, 3:4 (July 1851), p. 563. Charles Pickering was a naturalist on the expedition; see also his *The Gliddon Mummy-Case in the Museum of the Smithsonian Institution* (Washington, DC: Smithsonian Contributions to Knowledge, 1867).

43. Stanton, *The Leopard's Spots*, p. 102.

44. Ibid., p. 33.

45. Morton, cited in ibid., p. 33.

46. J. Nott and G. R. Gliddon, *Types of Mankind: or Ethnological Researches, based upon the Ancient Monuments, Paintings, Sculptures, and Crania of Races, and upon their Natural, Geographical, Philological and Biblical History: Illustrated by Selections from the Unedited*

Papers of Samuel George Morton and by Additional Contributions from Prof. L. Agassiz, LL.D., W. Usher, M.D., and Prof. H. S. Patterson, M.D. (Philadelphia, PA: Lippincott Gramoo & Co.; and London: Trubner & Co., 1855), p. xxxiii.

47. R. W. Emerson, 'Experience' (1848), in *Essays by Ralph Waldo Emerson: Second Series* (1848; Philadelphia, PA: David McKay, 1890), pp. 61–2. At other times, Emerson was given grudgingly to accept the scientific results of the phrenologists.

48. G. W. F. Hegel, *Hegel's Phenomenology of Spirit* (1807), ed. A. V. Miller and J. N. Findlay (New York: Oxford University Press, 1977), pp. 185, 200.

49. Ibid., p. 203.

50. Ibid., pp. 203–5.

51. H. S. Patterson, 'Memoir of the Life and Scientific Labors of Samuel George Morton', in Nott and Gliddon, *Types of Mankind*, pp. xvii–lvii, on p. xxxiii.

52. Others did not take this bait. They saw this progressive movement from the surface of the body to its base as revealing no real depth because the very act of revealing all immediately deprives them of depth. Everything becomes surface, Emerson says in his essay 'Experience', which brings him back to his critique of phrenology as deterministic.

53. Savage and Wyman 'Notice of the External Characteristics and Habits of Troglodytes Gorilla'.

54. Gould, *The Mismeasure of Man*, pp. 100–1.

55. S. G. Morton, *Brief Remarks on the Diversities of the Human Species* (Philadelphia, PA: Merrihew & Thompson, 1842), p. 133.

56. Stanton, *The Leopard's Spots*, p. 7.

57. Squire, 'Announcement of Morton's Death'. Patterson amplifies these views: 'The reception of Morton's *Crania Americana* by the learned was all he could have desired. Everywhere it received the warmest commendations. The following extract, from a notice in the *London Medico-Chirurgical Review* for October 1840, will show the tone of the British scientific press: 'Dr. Morton's method and illustrations in eliciting the elements of his magnificent Craniography, are admirably concise, without being the less instructively comprehensive. His work constitutes, and will ever be highly appreciated as constituting an exquisite treasury of facts, well adapted, in all respects, to establish permanent organic principles in the natural history of man'. 'Here we finish our account of Dr. Morton's American Cranioscopy; and by its extent and copiousness, our article will show how highly we have appreciated his classical production. We have studied his views with attention, and examined his doctrines with fairness; and with perfect sincerity in rising from a task which has afforded unusual gratification, we rejoice in ranking his "*Crania Americana*" in the highest class of transatlantic literature, foreseeing distinctly that the book will ensure for its author the well-earned meed of a Caucasian reputation.' Patterson, 'Memoir of the Life and Scientific Labors of Samuel George Morton', p. xxxiv.

58. Stanton, *The Leopard's Spots*, pp. 90–2, 144.

59. Pickering, *The Gliddon Mummy-Case*, p. 1.

60. Such relationships were not unusual at the time because there was not so great a distinction between a scientist and amateur in the nineteenth century. Thoreau was greatly happy to have supplied specimens to Agassiz, and one of those involved in getting a specimen of turtle to Agassiz alive related his great joy at having been mentioned in a footnote to one of Agassiz's immortal works. The relationship not only gave the amateur a little note, but served also as a means to be remembered by posterity.

61. 'The unrolling of mummies has become a kind of mania. Who can forget the mummy maid unrolled by Gliddon last year in Boston, which turned out to be a man. Mr. Glid-

don, we see, recently unrolled another one at New Orleans, which was a female, and maintained its sex.' Anon., 'Unrolling the Mummy of a Bishop', *Scientific American*, 7:27 (20 March 1852), pp. 209–16, on p. 216.

62. E. A. Poe, 'Some Words with a Mummy', in *The Complete Tales and Poems of Edgar Allan Poe*, ed. H. Allen (New York: Modern Library, 1938), pp. 535–48, on p. 543.

63. Ibid., p. 544.

64. Swift Buckingham was a founder of the *Calcutta Journal* and the *Athenaeum* (later merging into the *Nation* and later with the *New Statesman*), world traveller and writer. He designed a model city based upon the geometric layout of a pyramid and hierarchical social order in his *National Evils and Practical Remedies, with a Plan of a Model Town* (London: Peter Jackson, 1849).

65. Poe, 'Some Words with a Mummy', pp. 544, 546.

66. See B. R. Polin, 'Poe's Literary Use of "Oppodeldoc" and other Patent Medicines', *Poe Studies*, 4:2 (1971), pp. 30–2.

67. In June 1849, Poe was held at Moyamensing prison while awaiting trial for public drunkenness. On cemeteries, see N. B. Miller, '"Embalmed in Egyptian Ethnology": Racial Discourse and Antebellum Interest in the Ancient Near East', paper presented at the Annual Meeting of the American Studies Association, Atlanta, GA (November, 2004).

68. See also S. A. Cartwright, 'Diseases and Peculiarities of the Negro Race', in J. DeBow (ed.), *DeBow's Review Southern and Western States* (New Orleans, LA, and New York: AMS Press, 1851), pp. 315–23. This article famously proposed 'drapetomania [*Dysaesthesia aethiopica* or "rascality"], or the disease causing negroes to run away ... unknown to our medical authorities, although its diagnostic symptom, the absconding from service, is well known to our planters and overseers ... is as much a disease of the mind as any other species of mental alienation, and much more curable, as a general rule. With the advantages of proper medical advice, strictly followed, this troublesome practice that many negroes have of running away, can be almost entirely prevented, although the slaves be located on the borders of a free state, within a stone's throw of the abolitionists' (pp. 322–3).

69. S. G. Morton, *Hybridity in Animals and Plants, Considered in Reference to the Question of the Unity of the Human Species* (New Haven, CT: B. L. Hamlen, 1847), p. 210.

70. Prichard, cited in ibid., p. 210.

71. Morton, *Hybridity in Animals and Plants*, p. 212.

72. Ibid., p. 210.

73. Ibid., p. 211.

74. Ibid., p. 212.

75. Morton, cited in B. B. Minor, 'A Sketch of the Progress of Archaeological Science in America', *Southern Literary Messenger*, 11:7 (July 1845), pp. 420–33, on p. 432.

76. Ibid., p. 431.

77. Ibid., pp. 432, 433.

78. Ibid., p. 433.

79. Morton, cited in Nott and Gliddon, *Types of Mankind*, p. 304.

80. Ibid., p. 304.

81. Cuvier in later writings raises the possibility that there might be regional revolutions in between the larger global catastrophes.

82. See Livingstone, *Adam's Ancestors* for a fuller account of this debate on the 'pre-Adamic races'.

83. Morton, cited in Nott and Gliddon, *Types of Mankind*, p. 397.

84. Ibid., pp. 397–8.
85. Nott and Gliddon, *Types of Mankind*, p. xx.
86. Haven, *Archaeology of the United States*, p. 83.
87. *DeBow's Review, Agricultural, Commercial, Industrial Progress and Resources*, 9:2 (August 1850), p. 295.
88. Lurie correctly understands that the American preoccupation with the monogenesis/ polygenesis dispute resulted in 'the incorporation into European writings of a large body of observations from the New World relating to Indians, Negroes, and animal life'. E. Lurie, 'Louis Agassiz and the Races of Man', *Isis*, 45:3 (September 1954), pp. 227–42, on p. 228. The names of Wells, Morton, Nott, Gliddon, Bachman, Audubon and later Agassiz were not the designations of peripheral figures but key referents in the most intensive and important scientific and social question of the day. It was not, however, a sort of contest between the New and Old Worlds, but the integration of a particular scientific ideology deployed in each location. Edinburgh, Philadelphia and Charleston were centres for the study of human diversity. Prichard and Morton both studied there. Prichard's 1808 dissertation *De humani generis varietate* marked the start of his own enduring research into human variety and was often noted by the naturalists of the American School.
89. Agassiz, cited in Gould, *The Mismeasure of Man*, pp. 76–7.
90. Stanton, *The Leopard's Spots*, p. 102, quoting Agassiz's friend Jules Marcau in 1896.
91. Lurie, 'Louis Agassiz and the Races of Man', p. 229. For a more complete argument, see Lurie's monograph *Nature and the American Mind: Louis Agassiz and the Culture of Science* (New York: Science History Publications, 1974).
92. Lurie, 'Louis Agassiz and the Races of Man', p. 229.
93. Junior League of Charleston, *Charleston Receipts* (Charleston, SC: Walker Evans & Cogswell Co., 1950), recipe for soused fish. 'Receipts [i.e. recipes] to designate time-honored dishes according to ancestral wishes'.
94. J. J. Audubon to Victor Audubon, 24 December 1833, in F. H. Herrick, *Audubon the Naturalist: A History of his Life and Times*, 2 vols (New York: Appleton, 1917), vol. 2, p. 62.
95. Audubon's painting of the long-billed curlew has Charleston in the background as a tribute to the Bachman family.
96. 'Among early American writers Samuel G. Morton based his conclusion upon a careful investigation of racial types. His views were largely influenced by his interest in the question of polygenism or monogenism, which dominated the minds of the period. He reached the conclusion that there must have been a multiple origin of human races, and claimed that the distinguishing characteristics of races were intimately associated with their physical build ... Morton was followed by a number of writers whose viewers were colored by their endeavor to defend slavery as an institution. To them the problem of polygeny and monogeny was important particularly because the distinct origin and permanence of the type of the Negro seemed to justify his enslavement. The most important writings of this group are those of J. C. Nott and George R. Gliddon. Nott in his introduction to the "the grand problem ... is that which involves the *common origin* of races; for upon the latter deduction hangs not only certain religious dogmas, but the more practical question of the equality and perfectibility or races– we say more practical question," because, while Almighty Power, on the one hand, is not responsible to Man for the distinct origin of human races, these, on the other, are accountable to Him for the manner in which their delegated power is used towards each other. Whether an

original diversity of races be admitted or not, the *permanence* of existing physical types will not be questioned by any Archaeologist or Naturalist of the present day. Nor, by such competent arbitrators, can the consequent permanence of moral and intellectual particularities of types be denied. The intellectual man is inseparable from the physical man; and the nature of the one cannot be altered without corresponding change in the other.' F. Boas, *The Mind of Primitive Man, Revised Edition* (1935; New York: Free Press, 1963), pp. 34–5.

97. G. Jones, 'The Social History of Darwin's *Descent of Man*', *Economy and Society*, 7:1 (1978), pp. 1–23, on p. 5.

98. T. Blake (ed.), *Charleston, South Carolina Slaveholders of 1860*, at http://freepages.genealogy.rootsweb.com/~ajac [accessed 13 February 2009].

99. At www.Stjohnscharleston.org [accessed January 2009].

100. J. Bachman, 'The Duty of the Planter to his Family, to Society, and his Country', *Charleston Daily Courier*, 25 March 1862.

101. Anon., 'The Original Unity of the Human Race', p. 542.

102. *Journal of the Anthropological Institute of New York*, 1:1 (1871), p. 96.

103. A. Johnson, 'Race Improvement by Control of Defectives (Negative Eugenics)', *Annals of the American Academy of Political Science and Social Science*, 34:1 (July 1909), pp. 22–9, on p. 23.

3 Darwin in Context

1. Of Smith, Darwin said he simply did not know of his work, as it had appeared in an appendix to a treatise on timber for ship-building. As for others, Darwin wrote a brief summary of his predecessors for inclusion in later editions of the *Origin of Species*, saying that his haste finally to publish his work after ten years had resulted in his leaving this task for later. Darwin's account has been questioned repeatedly and upheld just as often.

2. P. Horan, in C. Darwin, *The Origin of Species, Complete and Fully Illustrated* (1859), ed. P. Horan (New York: Modern Library, 1979), p. viii.

3. Gould, *Ontogeny and Phylogeny*, pp. 28–32.

4. Ibid., p. 35. Theodosius Dobzhansky pointed out that going backward only a few generations would reduce the homunculus to the size of a sub-atomic particle. See *Mankind Evolving: The Evolution of the Human Species* (New Haven, CT: Yale University Press, 1962), p. 24.

5. N. Wade, 'A Family Feud over Mendel's Manuscript on the Laws of Heredity', *New York Times*, 1 June 2010.

6. G. Canguilhem, *Knowledge of Life* (1965), ed. P. Marrati and T. Meyers (New York: Fordham University Press, 2008), pp. 104–7.

7. C. Darwin, *The Descent of Man* (New York: Modern Library, 1995), p. 181.

8. Darwin, *The Origin of Species*, p. 460.

9. W. Benjamin, 'The Work of Art in the Age of Mechanical Reproduction' (1936), at http://www.marxists.org/reference/subject/philosophy/works/ge/benjamin.htm [accessed 2 June 2010]. See also M. Horkheimer, 'Revolt of Nature', in *Eclipse of Reason* (1947; New York: Continuum, 1974), pp. 92–127.

10. Lyell, *Scientific Journals on the Species Question*, p. 410.

11. Ibid., p. 410.

12. Ibid., pp. 328, 411. Darwin often put his work squarely alongside Lyell's.

13. Ibid., p. 56.

14. With infinite variety comes the possibility that anyone could be a slave as opposed to the fixity and specific nature of the slave's 'mind, body and soul'.
15. It is not too far from this to the positing, as Marx was at this same time, of society itself as an immense factory.
16. Darwin, *The Origin of Species*, p. 171. It is always interesting to note how modest and withdrawn – often literally into the tranquillity of his garden – Darwin was, leaving the theorists who drew upon him the task of proclaiming the importance of the new science.
17. F. Jacob, *The Logic of Life: A History of Heredity* (New York: Pantheon Books, 1973), p. 165.
18. Darwin *The Origin of Species*, pp. 120–1. See also S. J. Hyman, *Tangled Bank: Darwin, Marx, Frazer and Freud as Imaginative Writers* (New York, Atheneum, 1962), p. 33. With the image of the tangled bank, reminiscent of Shakespearean lyric, Darwin embraces all the rich complexity of life. The image of the Great Chain of Life is ordered, hierarchic, and static, essentially medieval; the great tree of life is ordered, hierarchic, but dynamic and competitive, a Renaissance vision; but the great Tangled Bank of life is disordered, democratic, and subtly interdependent as well as competitive, essentially a modern vision.
19. 'All persons who are conversant with the present state of Zoology must be aware of the great detriment which the science sustains from the vagueness and uncertainty of its nomenclature. We do not here refer to those diversities of language which arise from the various methods of classification adopted by different authors, and which are unavoidable in the present state of our knowledge. So long as naturalists differ in the views which they are disposed to take of the natural affinities of animals there will always be diversities of classification, and the only way to arrive at the true system of nature is to allow perfect liberty to systematists in this respect. But the evil complained of is of a different character. It consists in this, that when naturalists are agreed as to the characters and limits of an individual group or species, they still disagree in the appellations by which they distinguish it. A genus is often designated by three or four, and a species by twice that number of precisely equivalent synonyms; and in the absence of any rule on the subject, the naturalist is wholly at a loss what nomenclature to adopt. The consequence is, that the so-called commonwealth of science is becoming daily divided into independent states, kept asunder by diversities of language as well as by geographical limits. If an English zoologist, for example, visits the museums and converses with the professors of France, he finds that their scientific language is almost as foreign to him as their vernacular. Almost every specimen which he examines is labeled by a title which is unknown to him, and he feels that nothing short of a continued residence in that country can make him conversant with her science. If he proceeds thence to Germany or Russia, he is again at a loss: bewildered everywhere amidst the confusion of nomenclature, he returns in despair to his own country and to the museums and books to which he is accustomed. / If these diversities of scientific language were as deeply rooted as the vernacular tongues of each country, it would of course be hopeless to think of remedying them; but happily this is not the case. The language of science is in the mouths of comparatively few, and these few, though scattered over distant lands, are in habits of frequent and friendly intercourse with each other. All that is wanted then is, that some plain and simple regulations, founded on justice and sound reason, should be drawn up by a competent body of persons, and then be extensively distributed throughout the zoological world.' C. Darwin, Prof. Henslow, L. Jenyns, W. Ogilby, J. Phillips, Richardson, H. E. Strickland (reporter) and J. O. Westwood, 'Report of a Committee appointed "to Consider of the

Rules by which the Nomenclature of Zoology may be Established on a Uniform and Permanent Basis"', *Report of the British Association for the Advancement of Science for 1842* (London, 1843), pp. 106–7.

20. Singer, *A History of Biology*, p. 386.

21. See J. C. Greene, *The Death of Adam: Evolution and Its Impact On Western Thought* (Ames, IA: Iowa State University Press, 1996).

22. These same questions arise in our own day now that our closeness to other primates has been well established by the genome project and recognized recently by the Spanish parliament, and hopefully furthered by other nations as well.

23. W. Coleman, *Biology in the Nineteenth Century: Problems of Form, Function, and Transformation* (New York: John Wiley & Sons, 1971).

24. Wrightman, *The Growth of Scientific Ideas*, p. 414.

25. 'Natural Science will in time subsume the science of man, just as the science of man will subsume natural science. There will be *one* science ... Man is the immediate object of natural science ... But *nature* is the immediate object of the *science of man*. Man's first object – man – is nature, sense perception; and the particular sensuous human powers, since they can find objective realization only in *natural* objects, can find self-knowledge only in the science of nature in general ... the *social* reality of nature and *human* natural science or the *natural science of man* are identical expressions ... the creation of the earth received a heavy blow from the science of geogeny, i.e., the science which depicts the formation of the earth, its coming to be, as a process of self-generation. *Generatio æquivoca* is the only practical refutation of the theory of creation ... If you ask about the creation of nature and of man, then you are abstracting from nature and from man ... Give up your abstraction, and you will give up your question.' K. Marx, 'Economic and Social Manuscripts', in *Early Writings*, ed. R. Livingstone and G. Benton (New York: Penguin, 1974), pp. 279–400, on pp. 355–7.

26. See Canguilhem, *Ideology and Rationality* for a fuller discussion. Much has been made of the reputed desire of Marx to dedicate *Capital* to Darwin, but the relation of Darwin to Marx cannot be based on the dedication of *Capital*, but on the revolutionary aspects of their work. It is the demand for a transvaluation of all values (biological, natural, economic, ethical, etc.) and the imperative for interpretation that places Marx, Darwin, Emerson, Nietzsche and Freud together. They were in some respects the last gasp of the revolutionary bourgeoisie. The questioning of value and values relates these authors to each other. Marx recognized in Darwin not a kindred philosophy, but a kindred critique of philosophy, a critique that transcended philosophy to become a rational knowledge and experience of nature. See Jones, 'The Social History of Darwin's *Descent of Man*'.

27. Darwin, cited in Jacob, *The Logic of Life*, p. 165.

28. See Hyman, *Tangled Bank*.

29. Darwin to Karl Marx, 1 October 1873, in Darwin Correspondence Project Database, at http://www.darwinproject.ac.uk/darwinletters/calendar/entry-9080.html [accessed 25 May 2010].

30. K. Marx, *Grundrisse: Foundations of the Critique of Political Economy* (New York: Penguin, 1973), p. 1.

31. See also Francis Darwin in his preface to *The Life and Letters of Charles Darwin, including an Autobiographical Chapter*, ed. F. Darwin, 3 vols (London: John Murray, 1887).

32. Darwin, Notebook D [Transmutation of Species], 7–10 (1838), in Complete Works of Charles Darwin Online, at http://darwin-online.org.uk/content/record?itemID=CUL-DAR123 [accessed 26 July 2010].

33. Darwin, 'Autobiography', in *The Life and Letters,* vol. 1, pp. 67–9.
34. G. de Beer, in *Darwin: A Norton Critical Edition,* ed. P. Appleman, 3rd edn (New York: W. W. Norton, 2001), pp. 9, 5.
35. Ibid., p. 9.
36. Darwin, in *Darwin: A Norton Critical Edition,* p. 57.
37. Darwin, *The Origin of Species,* p. 117.
38. Darwin, *The Descent of Man,* p. 632.
39. Martin Naumann, personal communication, 1981.
40. Darwin to H. Spencer, 31 October 1873, in *More Letters of Charles Darwin,* ed. F. Darwin and A. C. Seward, 2 vols (London: John Murray, 1903), vol. 2, p. 351; Darwin to J. D. Hooker, Down, 15 January 1861, in *The Life and Letters,* vol. 2, p. 375.
41. H. Spencer to C. Darwin, in Lyell, *Scientific Journals on the Species Question,* p. 353.
42. J. Fiske, *Outlines of Cosmic Philosophy: Based on the Doctrine of Evolution, with Criticisms on the Positive Philosophy* (New York: Houghton, Mifflin & Co., 1916).
43. Referring to H. Spencer, 'Progress: Its Law and Cause', in *Essays, Scientific, Political, and Speculative* (London, 1863), pp. 8–62.
44. Lyell, *Scientific Journals on the Species Question,* p. 353.
45. M. Hawkins, for example, in *Social Darwinism in European and American Thought, 1860–1945: Nature as Model and Nature as Threat* (Cambridge: Cambridge University Press, 1998), spends a great deal of time arguing the distinction between Reformist Social Darwinism and hard Social Darwinism, particularly as a means to critique R. Hofstadter, *Social Darwinism in American Thought,* rev. edn (Boston, MA: Beacon Press, 1955).
46. Wrightman, *The Growth of Scientific Ideas,* p. 414. See P. A. Kropotkin, *Mutual Aid, a Factor of Evolution* (New York: Knopf, 1916); G. Nasmyth, *Social Progress and the Darwinian Theory; a Study of Force as a Factor in Human Relations,* intro. N. Angell (New York and London: G. P. Putnam's Sons, 1916); E. Haeckel, *The Evolution of Man: A Popular Exposition of the Principal Points of Human Ontogeny and Phylogeny* (Akron, OH: Werner Co., 1900); R. J. Herrnstein and C. Murray, *The Bell Curve: Intelligence and Class Structure in American Life* (New York: Simon & Schuster, 1996).
47. Wallace to Darwin, 2 July 1866, in *The Life and Letters,* vol. 3, pp. 58–9.
48. An idealist philosopher, it was said before Janet's *Causes Finales* was translated into English: 'Will there not be found in British science a man of eminence to fight the battle of good sense and of the facts, against the monstrous imagination of Darwin? If such a man comes out, he will find powerful assistants in our *Quatrefages,* our *Blanshard,* and our *Janet.* The book of this last one, on the Causes Finale, is really an event in science, and ought to have a large circulation among the educated classes abroad.' W. Affleck, cited in P. Janet, *The Materialism of the Present Day: A Critique of Dr. Buchner's System* (London: Williams & Norgate, 1867).
49. Darwin to Wallace, 5 July 1866, in *The Life and Letters,* vol. 3, pp. 46–7. The ironic humour of the last sentence is indicative of Darwin's critical remarks.
50. C. Darwin, *Journal of Researches into the Natural History and Geology of the Countries Visited during the Voyage of H.M.S. Beagle round the World, under the Command of Capt. Fitz Roy, R.N.,* 2nd edn (London: John Murray, 1845), p. 513.
51. 'Darwin had already encountered the phrase "struggle for existence" in a number of the works he had read, and he had used it in the 1844 Essay ... A fortunate change of ink after the first 18 folios of the manuscript of chapter five reveals significant details in its history. Darwin started writing this chapter under the title "On Natural Selection" and

only later decided to add "The struggle for existence" as the main theme. The original ink, now brown, is clearly distinguishable from the black of the later additions, notably in the title of the chapter, the added last sentence on folio 8: "This present chapter will be devoted to the Struggle for existence," and the slip of paper with the revised beginning of the direct discussion of this theme (fol. 9A). Although in the original brown ink version Darwin placed "War of nature" as an alternative to "Struggle of nature" as a rubric for this section, and began it in the Hobbesian vein, "all nature is at war," and although, through Erasmus Darwin he knew the even harsher Linnaean image of "One great slaughter-house the warring world!" he later changed his rubric to "struggle for existence". This he could interpret more broadly than war between organisms to include the physical environment as well: "A plant on the edge of a desert is often said to struggle for existence" (fol. 30A). *Charles Darwin's Natural Selection; being the Second Part of his Big Species Book written from 1836 to 1858*, ed. R. C. Stauffer (Cambridge: Cambridge University Press, 1975), p. 187.

52. Darwin, *The Origin of Species*, p. 63.
53. Ibid., pp. 58, 63.
54. Ibid., p. 69. If this is true then the tendency for population to increase geometrically and hence prompt the struggle for existence between individuals and populations must be seen as only part of what became a general ecological theory.
55. Ibid., p. 70. Anaximander 'And the source of coming-to-be for existing things is that into which destruction, too, happens, "according to necessity; for they pay penalty and retribution to each other for their injustice according to the assessment of Time"' (Simplicius, *Physics*, 24, 17, quoted in G. S. Kirk and J. E. Raven, *The Presocratic Philosophers; a Critical History with a Selection of Texts* (Cambridge: Cambridge University Press, 1957), p. 117.
56. Darwin, *The Origin of Species*, pp. 125–6.
57. N. Eldredge and S. J. Gould, 'Punctuated Equilibria: An Alternative to Phyletic Gradualism', in T. J. M. Schopf (ed.), *Models of Paleobiology* (San Francisco, CA: Freeman, Cooper & Co., 1972), pp. 82–115.
58. See Darwin's 'Posthumous Essay on Instinct', in G. J. Romanes, *Mental Evolution in Animals; with a Posthumous Essay on Instinct by Charles Darwin* (London: Kegan Paul, Trench & Co., 1883), pp. 355–84, where the list included 'the larvae of the Ichneumidae feeding within the live bodies of their prey, cats playing with mice, otters and cormorants with living fish, not as instincts specially given by the Creator, but as very small parts of one general law leading to the advance of all organic beings – multiply, vary, let the strongest live and the weakest die' (p. 384). Romanes says that this longer version of the chapter on instinct was 'suppressed for the sake of condensation', but it has a fundamentally different structure than does the published chapter with the late addition of the section on slave-making ants (p. 355).
59. Aristotle, *History of Animals*, I.1.
60. Darwin, *The Descent of Man*, p. 158.
61. H. D. Thoreau, *Walden and Other Writings*, ed. J. W. Krutch (New York: Bantam Books, 1962), pp. 274–8. This passage shows how important the study of natural history was to the formation of northern as well as southern intellectuals, perhaps one of the last vestiges of the Jeffersonian tradition. In his questionnaire for election to the Association for the Advancement of Science, an appointment he would later decline, Thoreau wrote that his special interest was in 'the Manners & Customs of the Indians of the Algonquin Group previous to contact with civilized man'. He describes himself as 'an observer

of nature generally and the character of my observations, so far as they are scientific, may be inferred from the fact that I am especially attracted by such books of science as White's *Selborne* and Humboldt's *Aspects of Nature*'. Thoreau was well acquainted with the work of the American School as well as having collected specimens for Agassiz and the collections at Harvard. See L. Wilson, 'Thoreau: Student of Anthropology', *American Anthropology*, 61:2 (1959), pp. 279–89; H. D. Thoreau, *Material Faith: Henry David Thoreau on Science*, ed. L. D. Walls (New York: Houghton Mifflin, 1999); and D. L. Sharpe, 'Turtle Eggs for Agassiz', in H. Sharpley, S. Rapport and H. Wright (eds), *A Treasury of Science* (New York, Harper Brothers, 1958), pp. 31–44. Thoreau's anti-slavery positions are well established.

62. In *Darwin's Scared Cause: How a Hatred of Slavery Shaped Darwin's Views on Human Evolution* (London: Allen Lane, 2009), Adrian Desmond and James Moore note the importance of Darwin's discussion of the slave-making ant. Just as is the case today, there was a common-sense emphasis on using the behaviour of social insects to explain human behaviour. They do not mention Thoreau's war of the ants, but they do mention other widely read accounts of the similarity between human and insect slavery. There is agreement that slavery and instinct was a key connection that Darwin found a means of severing. Desmond and Moore's intellectual biography is quite excellent and there is no reason to engage in polemics on any small points of disagreement. The purpose here is not to write such a biography of Darwin, but to discuss the context into which Darwin intervened and to highlight the role that slavery and scientific authority have played in shaping our understanding of the meaning of human variety. I wish to make clearer the formation of our scientific ideology of race.

63. Darwin, *The Origin of Species*, pp. 243, 244.

64. Ibid., p. 244.

65. 'Darwin, to A. G. Butler, 20 February [1879], in Darwin Correspondence Project Database, at http://www.darwinproject.ac.uk/entry-11889/ [accessed 9 June 2010]. Smith's book on ants, with Darwin's annotations, is in the Darwin Library, Edinburgh University Library.

66. *Charles Darwin's Natural Selection*, p. 386. See also Darwin, *The Origin of Species*, pp. 130–73.

67. Smith to Darwin, 10 November 1857, in Darwin Correspondence Project Database, at http://www.darwinproject.ac.uk/entry-2167/ [accessed 27 July 2010].

68. Smith to Darwin, 26 February 1858, in ibid., at http://www.darwinproject.ac.uk/entry-2226/ [accessed 27 July 2010].

69. Darwin to Smith, [before 9 March 1858, in ibid., at http://www.darwinproject.ac.uk/entry-2235a/ [accessed 27 July 2010].

70. Smith to Darwin, 30 April 1859, in ibid., at http://www.darwinproject.ac.uk/entry-2456/ [accessed 9 June 2010].

71. Darwin to W. E. Darwin, 26 April 1858, in ibid., at http://www.darwinproject.ac.uk/entry-2265/ [accessed 27 July 2010].

72. Darwin to J. D. Hooker, 6 May 1858, in *The Life and Letters*, vol. 2, p. 107.

73. Thoreau, *Walden and Other Writings*, p. 277.

74. Darwin, *The Origin of Species*, p. 245.

75. P. Huber, *The Natural History of Ants*, ed. J. R. Johnson (London: Longman, Hurst, Rees, Orme, & Brown, 1820), pp. 376, xv–xvii.

76. Darwin, *The Origin of Species*, p. 246.

77. Ibid., pp. 245–6.

78. Ibid., p. 247.

79. Anon., 'The Original Unity of the Human Race', p. 543.

80. 'Although in many cases it is most difficult to conjecture by what transitions an organ could have arrived at its present state; yet, considering that the proportion of living and known forms to the extinct and unknown is very small, I have been astonished how rarely an organ can be named, towards which no transitional grade is known to lead. The truth of this remark is indeed shown by that old canon in natural history of "Natura non facit saltum." We meet with this admission in the writings of almost every experienced naturalist; or, as Milne Edwards has well expressed it, nature is prodigal in variety, but niggard in innovation. Why, on the theory of Creation, should this be so? Why should all the parts and organs of many independent beings, each supposed to have been separately created for its proper place in nature, be so invariably linked together by graduated steps? Why should not Nature have taken a leap from structure to structure? On the theory of natural selection, we can clearly understand why she should not; for natural selection can act only by taking advantage of slight successive variations; she can never take a leap, but must advance by the shortest and slowest steps.' Darwin, *The Origin of Species*, p. 403.

81. Ibid., p. 460.

82. Morton, cited in Nott and Gliddon, *Types of Mankind*, p. 304.

83. Darwin, *The Origin of Species*, p. 460.

84. Ibid., p. 460.

85. Gould, *Ontogeny and Phylogeny*, pp. 28–9. Tobach argued that the homunculus has returned in modern genetics in the form of the 'cryptanthroparion' 'The prefix "crypt" referring to what is hidden in the little person (anthroparion); to the fact that the essential person is hidden in something; or that the person needs to be decoded in order to be understood. What is hidden in the cryptanthroparion? No one would deny the existence and the necessity of the material transmitted from progenitor to the progeny; that is, without the material being transmitted, there is no progeny. But, that is the sine qua non of all matter, animate and inanimate. Every phenomenon has a history. All matter is derived from other matter, though the form varies ... Hidden in the little person is the future of the little person.' E. Tobach, 'The Meaning of the Cryptanthroparion', in L. Ehrman, G. S. Omenn and E. Caspari (eds), *Genetics, Environment, and Behavior: Implications for Educational Policy* (New York: Academic Press, 1972), pp. 197–264, on p. 233.

86. The acceptance of evolution had little to do with Darwin, but Darwin certainly benefited from the general acceptance of preformist evolutionary theories.

87. See Darwin to Miles Joseph Berkeley, 7 April 1855, in Darwin Correspondence Project Database, at http://www.darwinproject.ac.uk/entry-1662/ [accessed 9 July 2010]. Darwin's interest in the first letter results from its focus on Gaertner's assertions on the 'lessened fertility of hybrids' despite his own table, 'yet in the text seems to disregard' examples of normal fertility in hybrids '– at least he often repeats (as at p. 102) that the fertility in hybrids never equals that of the pure species', while Gaerner's own table of results shows this not to be the case.

88. Dobzhansky, *Mankind Evolving*, pp. 27–8; see also T. Dobzhansky, 'On Genetics and Politics', *Social Education*, 32:2 (February 1968), pp. 142–6.

89. A. Weismann, cited in A. E. E. McKenzie (ed.), *Major Achievements of Science* (London: Cambridge University Press, 1960), p. 149.

90. A. Gray, *The Elements of Botany for Beginners and for Schools* (New York: American Book Co., 1887), pp. 176–7.
91. T. Huxley, 'Emancipation – Black and White', in *Lay Sermons, Addresses, and Reviews* (New York: Appleton & Co., 1903), pp. 17–23.
92. Darwin, cited in G. McConnaughey, 'Dawinism and Social Darwinism', *Osiris*, 9 (1950), pp. 397–412, on p. 399.
93. Darwin, *Journal of Researches*, p. 228.
94. *The Life and Letters*, vol. 2, p. 312; compare Freud's noting that he never experienced the 'oceanic feeling' his friend described as the knowledge and realization of the divine in *Civilization and its Discontents*, pp. 11–12.
95. Darwin to Miss C. Darwin, 22 May 1833, in *The Life and Letters*, vol. 2, p. 246.
96. *The Life and Letters*, vol. 2, p. 78.
97. Ibid., vol. 2, p. 150.
98. Darwin to Charles Lyell, Down, 25 August [1845], in ibid., vol. 2, pp. 341–2, on p. 341.
99. Darwin to Asa Gray, 5 June 1861, in ibid., vol. 2, pp. 373–4. Darwin dedicated his work on *The Different Forms of Flowers on Plants of the Same Species* (London: Appleton, 1896) to Asa Gray 'as a small tribute of respect and Affection'.
100. *More Letters of Charles Darwin*, vol. 2, p. 139.
101. *Journal of Researches*, p. vii.
102. 'FitzRoy's character was a singular one, with very many noble features; he was devoted to his duty, generous to a fault, bold, determined, and indomitably energetic, and an ardent friend to all under his sway. He would undertake any sort of trouble to assist those whom he thought deserved assistance. He was a handsome man, strikingly like a gentleman, with highly courteous manners ... The difficulty of living on good terms with a Captain of a Man of War is much increased by its being almost mutinous to answer him as one would answer anyone else; and by the awe in which he is held, or was held in my time, by all on board.' Darwin, *Journal of Researches*, p. 10.
103. C. Darwin, *The Autobiography of Charles Darwin 1809–1882*, ed. N. Barlow (London: Collins, 1958), p. 75.
104. *The Life and Letters*, vol. 2, pp. 50–1. Darwin's father said upon the return of the *Beagle* that the shape of his son's head had been changed by his time away.
105. Darwin to J. M. Herbert, Maldonado, Rio Plata, 2 June 1833, in *The Life and Letters*, vol. 2, p. 246.
106. R. Fitz-Roy, *Narrative of the Surveying Voyages of His Majesty's Ships Adventure and Beagle between the Years 1826 and 1836, describing their Examination of the Southern Shores of South America, and the Beagle's Circumnavigation of the Globe. Proceedings of the Second Expedition, 1831–36, under the Command of Captain Robert Fitz-Roy, R.N.* (London: Henry Colburn. 1839), p. 62.
107. Ibid., p. 62.
108. Darwin, *The Autobiography*, p. 76.
109. 'I saw Fitz-Roy only occasionally after our return home, for I was always afraid of unintentionally offending him, and did so once almost beyond mutual reconciliation. He was afterwards very indignant with me for having published so unorthodox a book (for he became very religious) as the *Origin of Species*. Towards the close of his life he was, as I fear, much impoverished, and this was largely due to his generosity. Anyhow, after his death a subscription was raised to pay his debts. His end was a melancholy one, namely suicide, exactly like that of his uncle, Lord Castlereagh, whom he resembled closely in

manner and appearance. His character was in several respects one of the most noble which I have ever known, though tarnished by grave blemishes.' Ibid., p. 139.

110. Darwin, *Journal of Researches*, p. 43.
111. Darwin to J. M. Herbert, Maldonado, Rio Plata, 2 June 1833, in *The Life and Letters*, vol. 2, p. 246.
112. Darwin to C. Darwin, Maladonado, Rio Plata, 22 May 1833, in ibid., vol. 2, pp. 244–6.
113. Darwin to J. S. Henslow, Rio de Janeiro, 18 May 1832 in ibid., vol. 2, p. 235.
114. F. Darwin, in *The Life and Letters*, vol. 2, p. 212.
115. Darwin to Charles Lyell, 25 August [1845], in ibid., vol. 2, pp. 341, 342.
116. Darwin, *Journal of Researches*, p. 23.
117. *Charles Darwin's Beagle Diary*, ed. R. D. Keynes (Cambridge: Cambridge University Press, 2001), 13 April, 1832, p. 56.
118. Darwin, *Journal of Researches*, p. 23.
119. Ibid., p. 24.
120. Darwin, *Beagle Diary*, 13 April 1832, p. 56.
121. Ibid., 13 April 1832, p. 56.
122. Darwin, *Journal of Researches*, p. 23.
123. Darwin, *Beagle Diary*, 16 April 1832, p. 58. In his entry of 15 April 1832, Darwin wrote that he and his party 'were obliged to have a black man to clear the way with a sword' (ibid., p. 58), while in the *Journal of Researches*, he writes that 'it was necessary that a man should go ahead with a sword' (p. 25).
124. Darwin, *Beagle Diary*, 14 April 1832, p. 58. The contrasts between the plantations recall Seneca's treatment of slavery in the *Letters*. Rather than slavery bringing civilized behaviour, slavery itself becomes civilized and civilization becomes slavery.
125. Darwin, *Journal of Researches*, p. 25.
126. Darwin, *Beagle Diary*, 15 April 1832, p. 58.
127. Darwin, *Journal of Researches*, p. 25.
128. Darwin, *Beagle Diary*, 17–18 April 1832, pp. 58–9; Darwin, *Journal of Researches*, p. 26.
129. Darwin, *Journal of Researches*, p. 26.
130. Ibid., p. 26.
131. Ibid., p. 500.
132. Darwin to Asa Gray, 17 September 1861, in *The Life and Letters*, vol. 2, pp. 272–4.
133. Darwin to J. D. Hooker, Down, 15 January [1861], in ibid., vol. 2, p. 375.
134. Darwin to Asa Gray, Down, 23 February [1863], in ibid., vol. 3, p. 23. See also Prof. J. E. Cairnes, *The Slave Power: an Attempt to Explain the Real Issues involved in the American Contest* (1862; New York: A. M. Kelley, 1968); and F. L. Olmsted, *The Cotton Kingdom; a Traveller's Observations on Cotton and Slavery in the American Slave States. Based upon Three Former Volumes of Journeys and Investigations by the Same Author*, ed. A. M. Schlesinger (New York: Knopf, 1953).
135. Darwin to Asa Gray, 17 September 1861, in *The Life and Letters*, vol. 2, pp. 376–8.
136. Darwin to Asa Gray, 5 June 1861, in ibid., vol. 2, pp. 272–4.
137. Darwin to Asa Gray, Down, 19 April 1865, in Darwin Correspondence Project Database, at http://www.darwinproject.ac.uk/entry-4467/ [accessed 27 July 2010].
138. Darwin, *Journal of Researches*, p. 523.
139. McConnaughey, 'Dawinism and Social Darwinism', p. 400; see also Hofstadter, *Social Darwinism in American Thought*, pp. 70–200.
140. Darwin to John Morley, Down, 24 March 1871, in *The Life and Letters*, vol. 3, pp. 324–5.
141. Darwin, *Beagle Diary*, 17–18 April 1832, pp. 58–9.

142. Darwin, *The Descent of Man*, p. 541. While Nott saw in *The Origin of Species* the repudiation of the polygenic theory – and he gracefully accepted defeat – he nonetheless appreciated it as a 'capital dig at the parsons'. Stanton, *The Leopard's Spots*, p. 175.
143. Darwin, *The Origin of Species*, p. 433.

Conclusion

1. Foucault, *The Order of Things*, pp. 127–8.
2. For the Enlightenment philosophers, 'the conclusion which alone can assure us of the truth of nature is not deductive, logical, or mathematical; it is an inference from the part to the whole. The essence of nature as a whole can be deciphered and determined only if we take the nature of man as our starting point. Accordingly, the physiology of man becomes the point of departure and the key for the study of nature. Mathematics and mathematical physics are banished from their central position and superseded, in the works of the founders of materialistic doctrine, by biology and general physiology. Lamettrie begins with medical observations; Holbach appeals especially to chemistry and the sciences of organic life. Diderot raised the objection to the philosophy of Condillac that it is not enough to take mere sensation as the fundamental element of all reality. Science must go beyond this limitation and show the cause of our sensations, which can nowhere be found but in our physical organization. Thus the basis of philosophy does not lie in the analysis of sensations but in natural history, in physiology, and in medicine.' E. Cassirer, *The Philosophy of the Enlightenment* (Boston, MA: Beacon Press, 1951), p. 66.
3. Darwin, *The Descent of Man*, p. 52.
4. R. E. L. Faris, 'Evolution and American Sociology', in S. Persons (ed.), *Evolutionary Thought in America* (New York: George Braziller, 1956), pp. 160–81, on p. 167.
5. Lyell, *Scientific Journals on the Species Question*, p. 56.
6. Freud granted that his view of the id might seem strikingly similar to Weismann's: 'Great interest attaches from our point of view to the treatment given the subject of life and the death of organisms in the writings of Weismann'. At first glance, Weismann's 'division of the living substance into mortal and immortal parts' might appear to be an 'unexpected analogy' or 'dynamic corollary' to Freud's view of the Pleasure Principle and the Death Instinct (Eros and Thantos). But any appearance of a significant conjunction is quickly dispelled by their views 'on the problem of death'. Freud seized on what he saw as a fundamental contradiction in Weismann's views. At times Weismann believed death to be universal and at other times argued that death only comes into existence with multicellular organisms. S. Freud, *Beyond the Pleasure Principle*, ed. J. Strachey (New York: W. W. Norton & Co., 1961), pp. 56–9.

WORKS CITED

Anderson, M., *The American Census: A Social History* (New Haven, CT: Yale University Press, 1990).

Anon., 'Table of Lunacy in the United States', *Hunt's Merchant's Magazine*, 8 (May 1843), pp. 460–2.

—, 'Reflections on the Census of 1840', *Southern Literary Messenger*, 9:6 (June 1843), pp. 340–52.

—, 'The Original Unity of the Human Race – Pickering, Bachman, Agassiz', *New England and Yale Review*, 8 (1850), pp. 542–85.

—, 'Unrolling the Mummy of a Bishop', *Scientific American*, 7:27 (1852), pp. 209–16.

—, 'The Aryan Question as it Stands Today: May Not the Original Home of the Indo-Germanic Peoples have been in Europe?', *New Englander and Yale Review*, 15:252 (1891), pp. 206–35.

Anon. ('Philanthropist'), 'Vital Statistics of Negroes and Mulattoes', *Boston Medical and Surgical Journal*, 17 (1842), pp. 168–70.

Aristotle, *The Politics of Aristotle*, ed. and trans. E. Barker (New York: Oxford University Press 1958).

Bachman, J., 'Unity of the Human Race', *Southern Quarterly Review*, 9:17 (January 1846), pp. 1–57.

—, 'The Duty of the Planter to his Family, to Society, and his Country', *Charleston Daily Courier*, 25 March 1862.

Benjamin, W., 'The Work of Art in the Age of Mechanical Reproduction' (1936), at http://www.marxists.org/reference/subject/philosophy/works/ge/benjamin.htm [accessed 2 June 2010].

Blake, T. (ed.), *Charleston, South Carolina Slaveholders of 1860*, at http://freepages.genealogy.rootsweb.com/~ajac [accessed 13 February 2009].

Blumenbach, J. F., *De generis humani varietate native*, 2nd edn (Gottingen, 1781).

—, *On the Natural Varieties of Mankind. De generis humani varietate nativa* (1776), ed. T. Bendyshe (New York: Bergman Publishers 1969).

—, *The Anthropological Treatises of Johann Friedrich Blumenbach*, trans. and ed. T. Bendyshe (Boston, MA: Longwood Press, 1978).

Boas, F., *The Mind of Primitive Man* (1935; New York: Free Press, 1963).

Brown, B. R., 'A City without Walls: Notes on Terror and Counter-Terrorism', *Situations: Project of the Radical Imagination*, 2:1 (2007), pp. 53–82.

Buck-Morss, S., 'Hegel and Haiti', *Critical Inquiry*, 26 (2000), pp. 821–67.

—, *Hegel, Haiti, and Universal History* (Pittsburgh, PA: University of Pittsburgh Press, 2009).

Buckingham, S., *National Evils and Practical Remedies, with a Plan of a Model Town* (London: Peter Jackson, 1849).

Buffon, G. L. Leclerc, comte de, *Natural History*, 10 vols (London: J. S. Barr, 1792).

Burnham, J. C. (ed.), *Science in America: Historical Selections* (New York: Holt, Rinehart & Winston, 1971).

Cairnes, J. E., *The Slave Power; its Character, Career and Probable Designs; being an Attempt to Explain the Real Issues involved in the American Contest* (1862; New York: A. M. Kelley, 1968).

Calhoun, J. C., *The Works of John C. Calhoun*, ed. R. K. Crallé, 6 vols (New York: D. Appleton, 1883).

Campbell, J., 'Folkloristic Commentary' (1944), in *The Complete Grimm's Fairy Tales* (New York: Partheon Fairy Tale and Folklore Library, 1972), pp. 833–64.

Canguilhem, G., *Ideology and Rationality in the History of the Life Sciences* (Cambridge, MA: MIT Press, 1988).

—, *Knowledge of Life* (1965), ed. P. Marrati and T. Meyers (New York: Fordham University Press, 2008).

Cartwright, S. A., 'Diseases and Peculiarities of the Negro Race', in J. De Bow (ed.), *DeBow's Review Southern and Western States* (New Orleans, LA, and New York: AMS Press, 1851), pp. 315–23.

Cassirer, E., *The Philosophy of the Enlightenment* (Boston, MA: Beacon Press, 1951).

Coleman, W., *Georges Cuvier, Zoologist: A Study in the History of Evolution Theory* (Cambridge, MA: Harvard University Press, 1964).

—, *Biology in the Nineteenth Century: Problems of Form, Function, and Transformation* (New York: John Wiley & Sons, 1971).

Colum, P., 'Introduction', in *The Complete Grimm's Fairy Tales* (New York: Partheon Fairy Tale and Folklore Library, 1972), pp. vii–xiv.

Conrad, J., *Heart of Darkness and the Secret Sharer*, ed. F. Walker (New York: Bantam Books, 1969).

—, *Selected Tales from Conrad*, ed. N. Stewart (London: Faber Paperbacks, 1977).

Coon, C. S., with E. E. Hunt, Jr, *The Living Races of Man* (New York: Alfred A. Knopf, 1966).

Cuvier, G., *A Discourse on the Revolutions of the Surface of the Globe, and the Changes Thereby Produced in the Animal Kingdom* (1817; Philadelphia, PA: Carey & Lea, 1831).

—, 'Mémoire sur l'ibis des anciens Egyptiens', in *Recherches sur les ossements fossiles de quadrupèdes: discours préliminaire* (Paris: GF-Flammarion, 1998), pp. 222–46.

Dana, J. D., 'Thoughts on Species', *American Journal of Science and Arts*, 24 (1857), pp. 305–16.

Darwin, C., *Journal of Researches into the Natural History and Geology of the Countries Visited during the Voyage of H.M.S. Beagle round the World, under the Command of Capt. Fitz Roy, R.N.*, 2nd edn (London: John Murray, 1845).

—, 'Posthumous Essay on Instinct', in Romanes, *Mental Evolution in Animals*, pp. 355–84.

—, *The Life and Letters of Charles Darwin, including an Autobiographical Chapter*, ed. F. Darwin, 3 vols (London: John Murray, 1887).

—, *The Different Forms of Flowers on Plants of the Same Species* (London: Appleton, 1896).

—, *More Letters of Charles Darwin*, ed. F. Darwin and A. C. Seward, 2 vols (London: John Murray, 1903).

—, *The Autobiography of Charles Darwin 1809–1882*, ed. N. Barlow (London: Collins, 1958).

—, *Charles Darwin's Natural Selection; being the Second Part of his Big Species Book written from 1836 to 1858*, ed. R. C. Stauffer (Cambridge: Cambridge University Press, 1975).

—, *The Origin of Species, Complete and Fully Illustrated* (1859), ed. P. Horan (New York: Modern Library, 1979).

—, *The Descent of Man* (New York: Modern Library, 1995).

—, *The Descent of Man*, 2nd edn (1874; New York: Prometheus Books, 1998).

—, *Darwin: A Norton Critical Edition*, ed. P. Appleman, 3rd edn (New York: W. W. Norton, 2001).

—, *Charles Darwin's Beagle Diary*, ed. R. D. Keynes (Cambridge: Cambridge University Press, 2001).

—, Complete Works of Darwin Online, ed. J. van Wyhe, at http://darwin-online.org.uk/.

—, Darwin Correspondence Project, at http://www.darwinproject.ac.uk/home.

Darwin, C., Prof. Henslow, L. Jenyns, W. Ogilby, J. Phillips, Richardson, H. E. Strickland (reporter) and J. O. Westwood, 'Report of a Committee appointed "to Consider of the Rules by which the Nomenclature of Zoology may be Established on a Uniform and Permanent Basis"', *Report of the British Association for the Advancement of Science for 1842* (London, 1843), pp. 106–7.

De Bow, J. D. B., 'The Hireling and the Slave (continued)', *Debow's Review, Agricultural, Commercial, Industrial Progress and Resources*, 18:4 (April 1855), pp. 459–62.

—, 'The Hireling and the Slave – Illustrations from Mr. Grayson's Poem', *Debow's Review, Agricultural, Commercial, Industrial Progress and Resources*, 21:3 (September 1856), pp. 248–56.

Dekay, J. E., 'Anniversary Address on the Progress on the Natural Sciences in the United States: Delivered before the Lyceum of Natural History of New York' (1826), in Burnham (ed.), *Science in America*, pp. 75–89.

Desmond, A., and J. Moore, *Darwin's Scared Cause: How a Hatred of Slavery Shaped Darwin's Views on Human Evolution* (London: Allen Lane, 2009).

Dobzhansky, T., *Mankind Evolving: The Evolution of the Human Species* (New Haven, CT: Yale University Press, 1962).

—, 'On Genetics and Politics', *Social Education*, 32:2 (February 1968), pp. 142–6.

Eldredge, N., and S. J. Gould, 'Punctuated Equilibria: An Alternative to Phyletic Gradualism', in T. J. M. Schopf (ed.), *Models of Paleobiology* (San Francisco, CA: Freeman, Cooper & Co., 1972), pp. 82–115.

Ellis, J., *The Social History of the Machine Gun* (London: Cressett Press, 1971).

Emerson, R. W., *Essays by Ralph Waldo Emerson: Second Series* (1848; Philadelphia, PA: David McKay, 1890).

Faris, R. E. L., 'Evolution and American Sociology', in S. Persons (ed.), *Evolutionary Thought in America* (New York: George Braziller, 1956), pp. 160–81.

Fiske, J., *Outlines of Cosmic Philosophy: Based on the Doctrine of Evolution, with Criticisms on the Positive Philosophy* (New York: Houghton, Mifflin & Co., 1916).

Fitz-Roy, R., *Narrative of the Surveying Voyages of His Majesty's Ships Adventure and Beagle between the Years 1826 and 1836, describing their Examination of the Southern Shores of South America, and the Beagle's Circumnavigation of the Globe. Proceedings of the Second Expedition, 1831–36, under the Command of Captain Robert Fitz-Roy, R.N.* (London: Henry Colburn, 1839).

Foucault, M., *The Order of Things: A History of the Human Sciences* (New York: Vintage Books, 1970).

Frank, T., 'Race Mixture in the Roman Empire', *American Historical Review*, 21 (1916), pp. 689–708.

Frazer, J. G., *Golden Bough: A Study in Comparative Religion*, 2 vols (London: Macmillan, 1890).

Freud, S., *Beyond the Pleasure Principle*, ed. J. Strachey (New York: W. W. Norton & Co., 1961).

—, *Civilization and its Discontents*, ed. J. Strachey (New York: W. W. Norton & Co., 1961).

—, *Future of an Illusion*, ed. J. Strachey (New York: W. W. Norton & Co., 1961).

Friedman, J. B., *The Monstrous Races in Medieval Art and Thought* (Cambridge, MA: Harvard University Press, 1981).

Gibbon, E., *The History of the Decline and Fall of the Roman Empire* (1783), ed. D. M. Milmam, M. Guizot and W. Smith (New York: Nottingham Society, 1845).

Gooch, G. P., *History and Historians in the Nineteenth Century* (London: Longmans, Green, & Co., 1920).

Gossett, T., *Race: The History of an Idea in America* (Dallas, TX: Southern Methodist University, 1963).

Gould, S. J., *Ontogeny and Phylogeny* (Cambridge, MA: Belknap Press of Harvard University Press, 1977).

—, 'Ladders and Cones: Constraining Evolution by Canonical Icons', in R. B. Silvers (ed.), *Hidden Histories of Science* (New York: New York Review Books, 1995), pp. 37–68.

—, 'The Man who Invented Natural History', Review of *Buffon* by Jacques Roger, trans. Sarah Lucille Bonnefoi, Cornell University Press, *New York Review of Books*, 45:16 (22 October 1998), pp. 83–6.

—, *The Mismeasure of Man*, 2nd edn (New York: W. W. Norton, 1999).

Gray, A., *The Elements of Botany for Beginners and for Schools* (New York: American Book Co., 1887).

Grayson, W. J., 'The Hireling and the Slave', *Debow's Review, Agricultural, Commercial, Industrial Progress and Resources*, 18:2 (February 1855), pp. 185–8; 19:2 (August 1855), pp. 208–18.

—, *Selected Poems by William J. Grayson*, comp. Mrs W. H. Armstrong (New York, Washington: Neale Publishing Co., 1907).

Greene, J. C., *The Death of Adam: Evolution and its Impact on Western Thought* (Ames, IA: Iowa State University Press, 1996).

Grimm, J., *Teutonic Mythology* (1844), ed. J. S. Stallybrass (New York: Dover Publications, 1966).

Gschwendtner, A. (dir.), *Der Menschen Forscher [The Anthropologist]* (1992).

Haeckel, E., *The Evolution of Man: A Popular Exposition of the Principal Points of Human Ontogeny and Phylogeny* (Akron, OH: Werner Co., 1900).

Hanno, *The Voyage of Hanno Translated, and Accompanied with the Greek Text*, trans. T. Falconer (London: T. Cadell, Jr & Davies, 1797).

Haven, S. F., *Archaeology of the United States; or, Sketches, Historical and Bibliographical, of the Progress of Information and Opinion respecting Vestiges of Antiquity in the United States* (Washington, DC: Smithsonian Institution and G. P. Putnam & Co., 1856).

Hawkins, M., *Social Darwinism in European and American Thought, 1860–1945: Nature as Model and Nature as Threat* (Cambridge: Cambridge University Press, 1998).

Hazen, C. D., *Modern European History* (New York: Henry Holt & Co., 1917).

Hegel, G. W. F., *Hegel's Philosophy of Right* (1821), ed. T. M. Knox (New York: Oxford University Press, 1952).

—, *The Philosophy of History* (1830–1), ed. C. J. Friedrich (New York: Dover Publications, 1956).

—, *Hegel's Phenomenology of Spirit* (1807), ed. A. V. Miller and J. N. Findlay (New York: Oxford University Press, 1977).

—, *Philosophy of Nature*, ¶262, at http://www.marxists.org/reference/archive/hegel/.

Heidegger, M., 'The Origin of the Work of Art' (1935), in A. Hofstadter (ed.), *Poetry, Language, Thought* (New York: Harper & Row, 1971), pp. 75–8.

—, *Early Greek Thinking* (San Francisco, CA: Harper & Row, 1984).

Herrick, F. H., *Audubon the Naturalist: A History of his Life and Times*, 2 vols (New York: Appleton, 1917).

Herrnstein, R. J., and C. Murray, *The Bell Curve: Intelligence and Class Structure in American Life* (New York: Simon & Schuster, 1996).

Hodgen, M. T., *Early Anthropology in the Sixteenth and Seventeenth Centuries* (Philadelphia, PA: University of Pennsylvania Press, 1964).

Hodgson, G., *The Gentleman from New York: Daniel Patrick Moynihan* (Boston, MA: Houghton Mifflin, 2000).

Hofstadter, R., *Social Darwinism in American Thought*, rev. edn (Boston, MA: Beacon Press, 1955).

Horkheimer, M., *Eclipse of Reason* (1947; New York: Continuum, 1974).

Horkheimer, M., and T. Adorno, *Dialectic of Enlightenment: Philosophical Fragments* (1947; Stanford, CA: Stanford University Press, 2002).

Huber, P., *The Natural History of Ants*, ed. J. R. Johnson (London: Longman, Hurst, Rees, Orme, & Brown, 1820).

Hudson, N., 'Hottentots and the Evolution of European Racism', *Journal of European Studies*, 34:4 (2004), pp. 308–32.

Huxley, T., *Lay Sermons, Addresses, and Reviews* (New York: Appleton & Co., 1903).

Hyman, S. J., *Tangled Bank: Darwin, Marx, Frazer and Freud as Imaginative Writers* (New York, Atheneum, 1962).

Jacob, F., *The Logic of Life: A History of Heredity* (New York: Pantheon Books, 1973).

Janet, P., *The Materialism of the Present Day: A Critique of Dr. Buchner's System* (London: Williams & Norgate, 1867).

Jarvis, E., 'Statistics of Insanity in the United States', *Boston Medical and Surgical Journal*, 27 (1842), pp. 116–21, 281–2.

—, 'Increase of Human Life, III', *Atlantic Monthly*, 24:146 (1869), pp. 711–19.

Jefferson, T., *The Jefferson Bible: The Life and Morals of Jesus of Nazareth extracted Textually from the Gospels* (St Louis, MO: N. D. Thompson Publishing Co., 1902).

—, *Notes on the State of Virginia* (1787), ed. W. Peden (New York: W. W. Norton & Co., 1954).

—, *Letters*, ed. M. D. Peterson (New York: Literary Classics of the United States, 1984).

Johnson, A., 'Race Improvement by Control of Defectives (Negative Eugenics)', *Annals of the American Academy of Political Science and Social Science*, 34:1 (July 1909), pp. 22–9.

Jones, G., 'The Social History of Darwin's *Descent of Man*', *Economy and Society*, 7:1 (February 1978), pp. 1–23.

Jones, W., *Eleven Discourses* (London: Thubner & Co., 1875).

Junior League of Charleston, *Charleston Receipts* (Charleston: Walker Evans & Cogswell Co., 1950).

Kirk, G. S., and J. E. Raven, *The Presocratic Philosophers; a Critical History with a Selection of Texts* (Cambridge: Cambridge University Press, 1957).

Koerner, L., *Linnaeus: Nature and Nation* (Cambridge, MA: Harvard University Press, 1999).

Kojeve, A., *Introduction to the Reading of Hegel: Lectures on the Phenomenology of Spirit*, assembled R. Queneau; ed. A. Bloom; trans. J. H. Nichols, Jr (Ithaca, NY: Cornell University Press, 1980).

Kropotkin, P. A., *Mutual Aid, a Factor of Evolution* (New York: Knopf, 1916).

Lamarck, J. P., *Philosophie zoologique: ou Exposition des considérations relative à l'histoire naturelle des animaux* (Paris: Dentu, 1809).

Larson, J. L., *Reason and Experience, the Representation of Natural Order in the Work of Carl von Linné* (Berkeley, CA: University of California Press, 1971).

Lawrence, W., *Lectures on Physiology, Zoology, and the Natural History of Man* (London: James Smith, 1823).

Leidy, J., 'On the Hair of a Hottentot Boy', *Journal and Proceedings of the Academy of Natural History of Philadelphia* (1847), p. 7.

Lester, P., 'L'anthropologie', in M. Daumas (ed.), *Histoire de la science* (Paris: Gallimard, 1957), p. 1367.

Liddell, H. G., and R. Scott, *A Lexicon Abridged from Liddell and Scott's Greek–English Lexicon* (Oxford: Clarendon Press, 1958).

Livingstone, D., *Adam's Ancestors: Race, Religion, and the Politics of Human Origins* (Baltimore, MD: Johns Hopkins University Press, 2008).

Lockwood, T., 'Is Natural Slavery Beneficial?', *Journal of the History of Philosophy*, 45:2 (2007), pp. 207–21.

Lovejoy, A. O., *The Great Chain of Being: A Study of the History of an Idea* (Cambridge, MA: Harvard University Press, 1936).

Lurie, E., 'Louis Agassiz and the Races of Man', *Isis*, 45:3 (September 1954), pp. 227–42.

—, *Nature and the American Mind: Louis Agassiz and the Culture of Science* (New York: Science History Publications, 1974).

Lyell, C., *Sir Charles Lyell's Scientific Journals on the Species Question* (1860), ed. L. G. Wilson (New Haven, CT: Yale University Press, 1970).

McConnaughey, G., 'Dawinism and Social Darwinism', *Osiris*, 9 (1950), pp. 397–412.

McKenzie, A. E. E. (ed.), *Major Achievements of Science* (London: Cambridge University Press, 1960).

Mallory, J. P., 'A History of the Indo-European Problem', *Journal of Indo-European Studies*, 1:1 (1973), pp. 21–65.

Macrobius, *The Saturnalia*, ed. P. V. Davies (New York: Columbia University Press, 1969).

Marx, K., *Grundrisse: Foundations of the Critique of Political Economy* (New York: Penguin, 1973).

—, 'Economic and Social Manuscripts', in *Early Writings*, ed. R. Livingstone and G. Benton (New York: Penguin, 1974), pp. 279–400.

—, 'The Viennese Ape Theatre in Berlin' (1837), in *Karl Marx, Frederick Engels: Collected Works*, 50 vols (London: Lawrence & Wishart; New York: International Publishers; Moscow: Progress Publishers, 1975–2004), vol. 1, p. 539.

—, *Capital*, 3 vols (New York: Penguin, 1976–81).

Meigs, J. A., *Catalogue of Human Crania, in the Collection of the Academy of Natural Sciences of Philadelphia, PA: Based upon the third edition of Dr. Morton's 'Catalogue of Skulls'* (Philadelphia, PA: J. B. Lippincott & Co., 1857).

Miller, N. B., '"Embalmed in Egyptian Ethnology": Racial Discourse and Antebellum Interest in the Ancient Near East', paper presented at the Annual Meeting of the American Studies Association, Atlanta, GA (November, 2004).

Minor, B. B., 'A Sketch of the Progress of Archaeological Science in America', *Southern Literary Messenger*, 11:7 (July 1845), pp. 420–33.

Montagu, A., *Man's Most Dangerous Myth: The Fallacy of Race*, 5th edn, revised and enlarged (New York: Oxford University Press, 1974).

Morton, S. G., *Crania Americana* (Philadelphia, PA: J. Dobson, 1839).

—, *Brief Remarks on the Diversities of the Human Species* (Philadelphia, PA: Merrihew & Thompson, 1842).

—, *Catalogue of Skulls of Man and the Inferior Animals in the Collection of Samuel George Morton* (Philadelphia, PA: F. Turner, 1844).

—, *Hybridity in Animals and Plants, Considered in Reference to the Question of the Unity of the Human Species* (New Haven, CT: B. L. Hamlen, 1847).

—, 'Some Observations on the Bushman Hottentot Boy', *Journal and Proceedings of the Academy of Natural History of Philadelphia* (1848), p. 5.

Mosse, G., *Nationalism and Sexuality: Respectability and Abnormal Sexuality in Modern Europe* (New York: Howard Fertig, 1985).

Muller, M., *Chips from a German Workshop*, 3 vols (New York: Charles Scribner & Co., 1872).

Murray, R. H., *Science and Scientists in the Nineteenth Century* (1925; New York: Macmillan Co., 1988).

Nasmyth, G., *Social Progress and the Darwinian Theory; a Study of Force as a Factor in Human Relations*, intro. N. Angell (New York and London: G. P. Putnam's Sons, 1916).

Neuberger, O., *Astronomy and History: Selected Essays* (New York: Springer-Verlag, 1983).

Nietzsche, F., *On The Genealogy of Morals; Ecce Homo (1887–1888)*, ed. W. Kaufmann and R. J. Hollingdale (New York: Vintage Books, 1969).

Norduskiold, E., *The History of Biology: A Survey* (New York: Alfred A. Knopf, 1921).

Nott, J. C., 'The Unity of the Human Race', *Southern Quarterly Review*, 9:17 (January 1846), pp. 1–57.

—, 'Yellow Fever Contrasted with Billious Fever – Reason for Believing it a Disease of Sui Generis ... Probably Insect or Animalcular Origin', *New Orleans Medical and Surgical Journal*, 4 (1848), pp. 563–601.

Nott, J., and G. R. Gliddon, *Types of Mankind: or Ethnological Researches, based upon the Ancient Monuments, Paintings, Sculptures, and Crania of Races, and upon their Natural, Geographical, Philological and Biblical History: Illustrated by Selections from the Unedited Papers of Samuel George Morton and by Additional Contributions from Prof. L. Agassiz,*

LL.D., W. Usher, M.D., and Prof. H. S. Patterson, M.D. (Philadelphia, PA: Lippincott Gramoo & Co.; and London: Trubner & Co., 1855).

Olmsted, F. L., *The Cotton Kingdom; a Traveller's Observations on Cotton and Slavery in the American Slave States. Based upon Three Former Volumes of Journeys and Investigations by the Same Author*, ed. A. M. Schlesinger (New York: Knopf, 1953).

Omi, M., and H. Winant, *Racial Formation in the United States: From the 1960s to the 1990s*, 2nd edn (New York: Routledge, 1994).

Ovington, J., *Voyage to Suratt* (London: Jacob Tonson, 1696).

Parrington, V. L., *The Romantic Revolution in America, 1800–1869* (New York: Harcourt Brace Jovanovich, 1954).

Parsons, J., *Remains of Japhet: Being Historical Enquiries into the Affinity and Origin of the European Languages* (London: for the author, 1767).

Patterson, H. S., 'Memoir of the Life and Scientific Labors of Samuel George Morton', in Nott and Gliddon, *Types of Mankind*, pp. xvii–lvii.

Pickering, C., *The Gliddon Mummy-Case in the Museum of the Smithsonian Institution* (Washington, DC: Smithsonian Contributions to Knowledge, 1867).

Pictet, A., *Les Origines indo-européennes, ou les aryas primitifs, essai de paleontologie linguistiqoe*, 2 vols (Paris: J. Cherbuliez, 1859–63).

Pliny, *Natural History*, ed. H. Rackham, 10 vols (Cambridge, MA: Loeb Classical Library, 1942–52).

Poe, E. A., *The Complete Tales and Poems of Edgar Allan Poe*, ed. H. Allen (New York: Modern Library, 1938).

Polin, B. R., 'Poe's Literary Use of "Oppodeldoc" and other Patent Medicines', *Poe Studies*, 4:2 (1971), pp. 30–2.

Pomeroy, S. B., *Xenophon Oeconomicus: A Social and Historical Commentary, with new English translation* (Oxford: Clarendon Press, 1994).

Pringles, H. A., *The Master Plan: Himmler's Scholars and the Holocaust* (New York: Hyperion, 2006).

Prokosch, E., *A Comparative Germanic Grammar*, William Dwight Whitney Linguistic Series (Philadelphia, PA: Linguistic Society of America and University of Pennsylvania, 1939).

Rask, R. K., *A Grammar of the Icelandic or Old Norse Tongue* (1811), ed. T. H. Markey (Amsterdam: Benjamins, 1976).

Renfrew, C., *Archaeology and Language: The Puzzle of Indo-European Origins* (New York: Cambridge University Press, 1987).

Ritvo, H. *The Animal Estates: The English and Other Creatures in the Victorian Age* (Cambridge, MA: Harvard University Press, 1987).

—, 'Zoological Nomenclature and the Empire of Victorian Science', in B. Lightman (ed.), *Victorian Science in Context* (Chicago, IL: University of Chicago Press, 1997), pp. 334–53.

Romanes, G. J., *Mental Evolution in Animals; with a Posthumous Essay on Instinct by Charles Darwin* (London: Kegan Paul, Trench & Co., 1883).

Rudwick, M. J. S., *Georges Cuvier, Fossil Bones, and Geological Catastrophes* (Chicago, IL: University of Chicago Press, 1997).

Rydberg, V., *Teutonic Mythology* (London: S. Sonnenschein & Co., 1889).

de Santillana, G., *Reflections on Men and Ideas* (Cambridge, MA: MIT Press, 1968).

de Saussure, F., *Course in General Linguistics* (Chicago, IL: Open Court, 1983).

Savage, T. S., and J. Wyman, 'Notice of the External Characteristics and Habits of Troglodytes Gorilla, a New Species of Orang from the Gaboon River', in Burnham (ed.), *Science in America*, pp. 115–27.

Sharpe, D. L., 'Turtle Eggs for Agassiz', in H. Sharpley, S. Rapport and H. Wright (eds), *A Treasury of Science* (New York: Harper Brothers, 1958), pp. 31–44.

Shirer, W. L., *The Rise and Fall of the Third Reich: A History of Nazi Germany* (New York: Fawcett Crest, 1950).

Singer, C., *A History of Biology: A General Introduction to the Study of Living Things* (New York: Henry Schumann, 1951).

Slotkin, J. S., *Readings in Early Anthropology: A Comprehensive Anthology of Pre-Scientific Writings on the Nature, Origin, History and Behavior of Man* (Chicago, IL: Aldine Publishing Co., 1965).

Smedley, A., *Race in North America: Origin and Evolution of a Worldview* (Boulder, CO: Westview Press, 1993).

Sorokin, P., *Social and Cultural Mobility* (New York: Free Press, 1959).

Spencer, H., *Essays, Scientific, Political, and Speculative* (London, 1863).

Spenser, F., 'Two Unpublished Essays on the Anthropology of North America by Benjamin Smith Barton', *Isis*, 68 (1977), pp. 567–73.

Squire, E. G., 'Announcement of Morton's Death to the Ethnological Society', *International Magazine of Literature, Art, and Science*, 3:4 (1851), p. 563.

Stanton, W. *The Leopard's Spots: Scientific Attitudes toward Race in America 1815–59* (Chicago, IL: University of Chicago Press, 1960).

Stephens, L. D., *Science, Race, and Religion in the American South: John Bachman and the Charleston Circle of Naturalists, 1815–1895* (Chapel Hill, NC: University of North Carolina, 2000).

Stocking, G. W., *Race, Culture, and Evolution: Essays in the History of Anthropology: With a New Preface* (Chicago, IL: University of Chicago Press, 1982).

Tacitus, *The Works of Tacitus*, ed. A. Murphy, 6 vols (Philadelphia, PA: Edward Earle, 1813).

Thoreau, H. D., *Walden and Other Writings*, ed. J. W. Krutch (New York: Bantam Books, 1962).

—, *Material Faith: Henry David Thoreau on Science*, ed. L. D. Walls (New York: Houghton Mifflin, 1999).

Tobach, E., 'The Meaning of the Cryptanthroparion', in L. Ehrman, G. S. Omenn and E. Caspari (eds), *Genetics, Environment, and Behavior: Implications for Educational Policy* (New York: Academic Press, 1972), pp. 197–264.

Toynbee, A., *A Study of History* (New York: Dell Publishing, 1969).

Wade, N., 'A Family Feud over Mendel's Manuscript on the Laws of Heredity', *New York Times*, 1 June 2010.

Webster, H., *Ancient History* (New York: D. C. Heath & Co., 1913).

Wilson, L., 'Thoreau: Student of Anthropology', *American Anthropology*, 61:2 (1959), pp. 279–89.

Wood, M., *Hitler's Search for the Holy Grail* (Mayavision International, 1998).

Wrightman, W. P. D., *The Growth of Scientific Ideas* (New Haven, CT: Yale University Press, 1951).

Yonge, C. M., *The Chosen People: A Compendium of Sacred and Church History for School-Children* (London: J. & C. Mozley, 1859).

INDEX